AF846661

Selected Topics in Statistical Inference

Manisha Pal · Bikas K. Sinha

Selected Topics in Statistical Inference

Theory and Applications

Springer

Manisha Pal
St. Xavier's University
Kolkata, India

Bikas K. Sinha
Indian Statistical Institute
Kolkata, India

ISBN 978-981-97-2591-5 ISBN 978-981-97-2592-2 (eBook)
https://doi.org/10.1007/978-981-97-2592-2

© The Editor(s) (if applicable) and The Author(s), under exclusive license to Springer Nature Singapore Pte Ltd. 2024

This work is subject to copyright. All rights are solely and exclusively licensed by the Publisher, whether the whole or part of the material is concerned, specifically the rights of translation, reprinting, reuse of illustrations, recitation, broadcasting, reproduction on microfilms or in any other physical way, and transmission or information storage and retrieval, electronic adaptation, computer software, or by similar or dissimilar methodology now known or hereafter developed.
The use of general descriptive names, registered names, trademarks, service marks, etc. in this publication does not imply, even in the absence of a specific statement, that such names are exempt from the relevant protective laws and regulations and therefore free for general use.
The publisher, the authors and the editors are safe to assume that the advice and information in this book are believed to be true and accurate at the date of publication. Neither the publisher nor the authors or the editors give a warranty, expressed or implied, with respect to the material contained herein or for any errors or omissions that may have been made. The publisher remains neutral with regard to jurisdictional claims in published maps and institutional affiliations.

This Springer imprint is published by the registered company Springer Nature Singapore Pte Ltd.
The registered company address is: 152 Beach Road, #21-01/04 Gateway East, Singapore 189721, Singapore

If disposing of this product, please recycle the paper.

In loving memory of my parents and my husband [Atul].

To my beloved daughters [Yashodhara and Jyotika] for their continued support.

—Manisha Pal

To our Research Collaborators—Beyond the Barriers and Boundaries—All over the Globe.

—Bikas K. Sinha and Manisha Pal

Preface

At one point of time, Prof. Manisha Pal [MP] acted as the co-ordinator of Research Advisory Committee for doctoral students in Statistics of Calcutta University [CU]. The students had to go through a series of courses—covering both theory and application areas in statistics. Professor Bikas K. Sinha [BKS]—an ex-student and an ex-faculty of the Department of Statistics, CU—was inducted as one of the instructors. For several years, BKS offered inference courses to the research scholars. MP also offered special lectures involving such topics as unbiased estimation of reliability based on samples from exponential distributions.

Both BKS and MP have developed special likings for some topics in Statistical Inference over long years of teaching in different universities and institutions. We got the opportunity to nurture our thoughts while addressing research students of Statistics in CU. That gave us the motivation to prepare Lecture Notes, with the intention to convert it into a Book entitled *Selected Topics in Statistical Inference.*

It is admittedly true that there are plenty of statistical inference books available in the market. We decided not to venture into writing another textbook. We thought of special topics—encompassing the areas covered by us while offering research courses to Statistics students. We have spent time and attention in conducting collaborative research on the topics included in the book for the last 40 years or so. We fervently hope that teachers, research scholars, and master's students will find the presentations stimulating and motivational.

While the manuscript was very much within our sight, one of us [BKS] had a chance encounter with Ms. Nupoor Singh of Springer and she wished to consider the same. We explained our plan about the contents of the book—mainly that it will not be a textbook in the usual sense. In due course, we volunteered to send her a few sample chapters, which she floated among a few subject experts. At the same time, we had also sent a few chapters to three of our research collaborators. The feedback turned out to be quite encouraging for us to go ahead and complete the task.

Eventually, we entered into an agreement with Springer Nature. We take pleasure in going for publication of the book whose contents have been carefully chosen as per our expertise and academic satisfaction.

BKS has a special liking and fascination for inference problems involving the Bernoulli parameter ‘p’ and its functions. This goes back to early 1970s when [Late] Prof. D. Basu used to address probabilists and statisticians at ISI, Kolkata. Basu’s penetrating thoughts and expositions opened up new vistas to all—young and old. Bimal [twin brother of Bikas] would be the first-hand listener sitting in the lecture rooms at ISI and he would transfer all that to BKS in at-home sessions. At that time BKS was a faculty at the Department of Statistics, CU. BKS got fascinated also after reading a short interesting paper by Gupta (1967) in Tokyo Annals. [Gupta: [Late] Prof. Milan K. Gupta of Presidency College, Kolkata]. Since then—over all these long 45+ years—BKS has, time and again, turned back to the mysterious world of Bernoulli trials. MP also got motivated to work in this area after attending a few lecture sessions of BKS. Chapter 2 exclusively deals with inference problems in binomial, trinomial, and tetranomial distributions involving such parameters as $[(p, q) \mid 0 < p, q < 1; p + q = 1]$ and $[(p, q, r) \mid 0 < p, q, r < 1; p + q + r = 1)]$. Our emphasis is on unbiased estimation of parameters/parametric functions.

Again, since the mid 1970s, BKS got attracted to finite population inference and started teaching advanced level courses and conducting/guiding research. At some point of time, among other research problems, he was fascinated by the problem of effective use of additional resources. Of course, the whole issue starts with a folklore: ‘The more the merrier’! This goes without saying under [SRSWOR (N, n), sample mean] strategy. There is more to it when we introduce general sampling designs and ask for extension of such designs with additional units. We have enjoyed unfolding the mysteries of effective use of additional resources and have presented the results in Chapter 3 for the readers.

Our experience shows that the so-called concept of sufficiency or of a sufficient statistic usually bothers both UG/PG students alike. They are told about the factorization theorem and its implications rather artificially. Upon reading Blackwell’s classic papers (1951, 1953) on ‘Comparison of Experiments’ and DeGroot’s paper (1966), BKS had the realization that there is more to this concept, and it can be dealt with quite elegantly. During the research course, BKS offered plenty of interesting examples from his own/joint work [BKS (1971), Dhungana et al. (2010a), Tiensuwan et al. (2010b, 2011)]. Chapter 4 grew out of these realizations and an urge to transmit all that among the researchers/readers at large.

Estimation of the size (N) of a finite population [of labeled units] is a standard problem in its most common framework. It is tacitly assumed that the population is ‘static’ in nature. [households, shops, hospitals, etc., are some examples]. There is a vast literature in this area—Vide Goodman (1953) and Seber (1982). Further references are De and Sengupta (2003) and Sengupta and De (1997). However, there are instances wherein we have to deal with what are identified as ‘Mobile Populations’. [tourists in resorts, visitors in shopping centers, etc.]. Another scenario arises when we have to deal with individuals suffering from a rare disease and direct identification/enumeration of such individuals is impossible. It is only through the relevant service providers [hospitals, nursing homes, etc.] that such individuals can be identified and taken into account. Estimating the size of the population in such uncommon cases is covered in Chapter 5.

MP has been working in the area of reliability for more than a decade. Uniformly minimum variance unbiased (UMVU) estimator of the reliability of a product, as well as stress-strength reliability, is not at all simple to deduce even in the case of one-parameter exponential distribution, and not much literature exists in this area. The situation becomes even more complex when the lifetime data is censored, or only the lifetimes of a few of the items put on life test are available. Chapter 6 gives an explicit narration of UMVU estimation of reliability in the one-parameter exponential set-up and also indicates interval estimation of the same.

Kolkata, India

Manisha Pal
Bikas K. Sinha

Acknowledgements

IIT Mumbai Conference ... followed by UniPune Conference ... Prof. Bikas K. Sinha (BKS) chanced to meet Ms. Nupoor Singh [NS] of Springer Nature at both the conferences. NS was interested to know if BKS had any text book material at hand/in mind for publication and that, if so, she would very much like to consider it. That was something any author would love to hear from a reputed publisher! BKS shared with her his plan towards publication of a manuscript in Statistical Inference with selected topics along with Prof. Manisha Pal. NS encouraged him to take up the project and send the manuscript to her. That is how we, the authors, got into this business of manuscript preparation/compilation—covering exclusively the topics of our liking.

As a gesture of our gratitude to Springer Nature [SN], we profusely thank NS for providing us with an opportunity for a successful collaboration. We fondly hope we have not disappointed her.

Thanks are also due to Prof. Bimal K. Sinha [twin brother of BKS and Presidential Research Professor of Statistics at UMBC, Maryland, USA]. He was the first and foremost reader of our statistical thoughts and marked it as an excellent venture!

Both of us have gone ahead to expose specific parts of the technical contents at national and international workshops/conferences, taking into account the valuable comments received from interested participants. Such sporadic comments shaped our style of presentation from a *dull show* to a *lively entertainment*! We wish to record our appreciation for the time and thought given by such enthusiasts.

Springer Nature Team deserves a special mention for handling our novice efforts [in manuscript typing/preparation] with extreme courtesy at all stages.

June 2023

Manisha Pal
Bikas K. Sinha

Contents

About the Authors

Manisha Pal is a retired professor of Statistics from the University of Calcutta, India and is currently a senior professor in the Department of Statistics, St. Xavier's University, Kolkata, India. She has been involved in fruitful research since 1982 and has more than 130 research publications in peer-reviewed journals, having collaborated with many researchers in India and abroad. Her co-authored book, *Optimal Mixture Experiments*, was published in Springer's prestigious *Lecture Notes in Statistics* book series. Her areas of research interest are inventory control, reliability inference, skewed distributions, mixture experiments and data analysis.

Bikas K. Sinha is a retired professor of Statistics from Indian Statistical Institute, Kolkata. He has contributed immensely to both statistical theory and applications. He has published a number of research monographs and textbooks with publishers of international repute, including Springer. With more than 110 research collaborators worldwide and more than 160 research publications in peer-reviewed journals, and three research monographs (co-authored) published by Springer in its *Lecture Notes in Statistics Series* (1989, 2002, 2014), his scholastic contributions have earned him international academic recognition in Statistics.

Chapter 1
Glimpses of the Book

1.1 Introduction

We intend to present certain finer aspects of topics related to statistical estimation—theory and applications. We have been guided by our pedagogical experiences while teaching statistical inference courses at different levels within India and outside as well. We have found that there are certain issues which are taken for granted without having closer look into them. Our primary aim is to discuss these issues—spanning over a number of teaching and research areas. Some of these issues may seem to be simple to start with. As we proceed, the readers will appreciate how existing knowledge gaps are fruitfully explained and covered with relevant theory and worked-out examples.

We have altogether 5 distinct topics—going from Chaps. 2 to 6. In each chapter, we start with an introduction and explanation of necessary background details before taking up the questions of relevance, followed by discussions. We have tried to make the presentations in a lucid style—highlighting the importance of the theoretical methods, followed by illustrative examples.

1.2 Chapter-Wise Description of Contents

In Chap. 2, we take up the age-old study in the domain of repeated Bernoulli trials with success probability [showing up a Head] as $p, 0 < p < 1$.

Here is an incidence to be narrated. BKS was a Faculty at ISI, Kolkata, and on one summer day, 4 students [2 from IIT, Kanpur, and 2 from IIT, Kharagpur] approached him with a request for 'Summer Training' for 6 weeks. They had recommendations from their teachers to meet BKS specifically. Naturally, BKS was very happy and he wanted to know the topics that would interest them. They left the choice of the topic(s) to BKS. BKS asked if they would be interested in knowing 'more' about repeated Bernoulli trials! The summer trainees started laughing—disbelieving that

© The Author(s), under exclusive license to Springer Nature Singapore Pte Ltd. 2024
M. Pal and B. K. Sinha, *Selected Topics in Statistical Inference*,
https://doi.org/10.1007/978-981-97-2592-2_1

there is still enough unknown material involving the Bernoulli/binomial distributions. BKS assured them: I will deliver 10 Lectures [each of 1 hour duration] and at the end—you will find to your utter surprise that all these were totally unheard of so far!!! And, BKS gave them the famous quiz—originally posed by D. Basu: Generate an event with probability $\sqrt{p}$. Their summer course journey for the next 6 weeks was centered around 'p' and its functions. This topic is predominantly discussed in Chap. 2. Generalizations of binomial to trinomial and then to tetranomial distributions have also been discussed with meaningful non-trivial estimation problems.

In Chap. 3, we take up yet another interesting study involving the sample size n in sampling from finite populations. It is implicitly taken for granted by all of us that an increase in the sample size n yields an increase in efficiency of the underlying estimator. This refers to the estimation of a finite population mean and use of the sample mean under $\text{SRSWR}(N, n)/\text{SRSWOR}(N, n)$ sampling schemes. A natural and yet pertinent question would be to examine if the same phenomenon holds for arbitrary Fixed Size (n) sampling designs—other than the usual $\text{SRSWR}/\text{SRSWOR}(N, n)$ sampling schemes. This question is simply posed. However, there are deep-rooted implications and consequences which are to be studied in a rather serious manner.

Another mind-blowing topic is the notion of 'Sufficiency' in statistical inference—covering both theory and applications. As students, we are taught about the 'Factorization Theorem' to identify the sufficient statistic(s) and proceed therefrom. It seems to be too mechanical, and further insights are mostly lacking. In Chap. 4, we propose to undertake a detailed study of this apparently mysterious topic and take up rigorous studies in various distributional settings.

Estimation of N, the unknown size of a finite population, is yet another interesting topic and we get exposure to the celebrated 'Catch-Recatch' technique in an elementary course in inference. However, there is much more to be told, studied, and researched in rather unusual complicated contexts. This is our topic in Chap. 5.

Reliability estimation based on a random sample from an exponential population is another interesting study, and we are exposed to the basic results in a course on reliability and life testing. How much we know is not very clear to most of us unless we face the deeper questions from practical perspectives. Let alone other life distributions, our study deals only with exponential distributions but under various restrictions on the nature of available data. This is our contribution in Chap. 6 for the readers.

We must hasten to add that all these chapter contents are very much to our liking, and we have dealt with these topics at various degrees [of depth] and at various levels for regular students and in refresher courses/research courses within India and elsewhere. We fondly hope the students/researchers/teachers will find our presentations and discussions stimulating and worthy of study. Our efforts will be rewarded if the readers pursue some of these topics rather seriously. We are sure they will find ample scope for unfolding the beauty of research in these areas.

Chapter 2
Sequential Binomial Estimation

Learning Objectives
Usually, Wald's Sequential Probability Ratio Test [SPRT] is a topic discussed in standard inference courses dealing with sequential analysis of data. On the other hand, (sequential) estimation of parameters in standard distributions is mostly left unaddressed or inadequately addressed!

Our aim is to discuss at length various aspects of the fascinating study of sequential estimation of Bernoulli parameter (p) and functions thereof; also, trinomial and tetranomial distributions will be considered in sequential framework.

The interested readers will find plenty of new ideas floating around. One of the authors [BKS] has been involved in this study since 1965 or so and has periodically visited this area of research with equal enthusiasm. Also, the other author [MP] has developed keen liking for this topic in the recent past.

2.1 Introduction

Consider the simplest and most elementary experiment of tossing a coin. This forms a Bernoulli trial with the two possible outcomes—head and tail, and respective probabilities, say, $P[\text{Head}] = p$ and $P[\text{Tail}] = 1 - p$, $0 < p < 1$. For estimating p, we consider 'repeated Bernoulli trials (RBTs)', with a pre-fixed finite number of trials, say n, which leads us to the (unique) minimum variance unbiased estimator of p, namely *the sample proportion of heads*. The earliest theory, pertaining to repeated Bernoulli trials, was developed about seventy years back to examine the scope of 'binomial sampling plans', providing unbiased estimators of the Bernoulli parameter 'p' and some functions of it.

According to Late Professor D. Basu, whatever we usually learn/teach in the framework of RBTs can be termed as being based on 'Enumeration/Computation' of probability. The real challenge lies in identifying events whose probabilities are

© The Author(s), under exclusive license to Springer Nature Singapore Pte Ltd. 2024

M. Pal and B. K. Sinha, *Selected Topics in Statistical Inference*,
https://doi.org/10.1007/978-981-97-2592-2_2

some specified and given functions of p. In this regard, the landmark question posed by him was—how can one generate an event with probability $\phi(p) = \sqrt{p}$, and subsequently—if possible, suggest an unbiased point estimator for $\phi(p)$. These are simply posed problems with far-reaching implications in the context of unbiased estimation of functions of p. Sinha and Banerjee (Sankhya, 1979) provided a non-trivial affirmative solution to the more general problem of estimating $\phi(p) = p^\alpha$, where α is a positive rational fraction.

In this chapter, we discuss in considerable details results pertaining to unbiased sequential estimation of 'p' and its functions, and also its generalizations for trinomial and tetranomial distributions. Before we start with the basic notations and nomenclature, we are tempted to mention two examples of $\phi(p)$, namely (i) $\phi(p) = 3.5pq$ and (ii) $\phi(p) = 12p^2q^2$, with not-so-immediate-solutions, toward generating the respective underlying events, which do exist but are not readily visible! In each case, we need to find a solution to the twin pair (n, E_n) where 'n' denotes the minimum number of Bernoulli trials (required to define the underlying event), and E_n is the underlying event for which $P[E_n] = \phi(p)$, $0 < p < 1$. Another related problem is to find the UMVUE of $\phi(p)$, say $\widehat{\phi}(p)$, which is also *proper* in the sense that $0 \le \widehat{\phi}(p) \le 1$ [since $\phi(p)$ is itself proper, i.e., $0 \le \phi(p) \le 1$ for $0 \le p \le 1$]. However, $\widehat{\phi}(p)$ is not necessarily so, irrespective of the choice of $\phi(p)$ and the *data-generating scheme* adopted! Examples of improper unbiased estimators are plenty, and the students/ researchers should be familiar with such examples in the context of the Bernoulli parameter 'p', toward working out proper unbiased estimators!

2.2 Notations and Nomenclature

Let $\{Z_i,\ i = 1, 2, \ldots\}$ denote an i.i.d sequence of Bernoulli variates resulting from repeated tosses of a coin (i.e., one and the same coin), with $P[Z_i = 1] = p$ and $P[Z_i = 0] = 1 - p = q$ (say). Any realization of this process can be exhibited as a lattice path in the (X, Y)-plane, where a particle moves one step from the origin to the right along the X-axis if the incoming observation is 0 [failure] and one step above along Y-axis if it is 1 [success]. Here we are assuming that the particle moves through integer co-ordinates in both the directions. A *stopping rule* can be defined as a sequence of functions ψ_k of $(Z_1, Z_2, \ldots, Z_k)$, where each ψ_k takes values 0 or 1. Given an integer 'k' and the realization $(z_1, z_2, \ldots, z_k)$, $\psi_k(z_1, z_2, \ldots, z_k) = 1$ indicates that we take one more observation, i.e., $(k + 1)$th observation and $\psi_k(z_1, z_2, \ldots, z_k) = 0$ indicates that we stop at this kth stage.

ALERT 1: We alert the readers of all levels about the uncommon nomenclature we are adopting in this chapter [upto a certain extent]. This refers to the representation of heads along the Y-axis and tails along the X-axis. This might create confusion from time to time in different contexts. This alert should be used in good sense.

Some Definitions

A point $\alpha = (x, y)$ is called a ***continuation point*** if there exists one sequence of realizations $(z_1, z_2, \ldots, z_{x+y})$ leading to α such that $\psi_j(z_1, z_2, \ldots, z_j) = 1\ \forall j \leq x+y$.

A point $\alpha = (x, y)$ is termed as a ***boundary point*** if there exists one sequence of realizations $(z_1, z_2, \ldots, z_{x+y})$ leading to α such that $\psi_j(z_1, z_2, \ldots, z_j) = 1\ \forall j < x+y$ and $\psi_{x+y}(z_1, z_2, \ldots, z_{x+y}) = 0$.

A point may be a boundary point or a continuation point depending on the path. This statement seems a bit puzzling. However, it will soon be made amply clear at an appropriate place.

A point is an ***accessible point*** if it is either a boundary point or a continuation point.

Points which are not accessible are regarded as ***inaccessible points***. In general terms, for any given collection of boundary points, there are usually, and trivially, an infinite collection of inaccessible points.

For any boundary point $\alpha = (x, y)$, $P(\alpha)$ denotes the probability of stopping at α and is given by

$$P(\alpha) = k(\alpha)p^y q^x, \text{ say}$$

where $k(\alpha)$ is the number of *accessible* paths from the origin to the point α.

A stopping rule yielding the boundary points, together with their probabilities $P(\alpha)$, is called a ***sampling plan*** P. It is also called a 'sequential Bernoulli sampling plan' since it is basically sequential in nature and it arises from Bernoulli trials. In Fig. 2.1, the boundary points are shown in red, and the free motion of the random particle is shown in blue.

The accessible/inaccessible points are not shown in Fig. 2.1 as those depend on the particular sampling plan being used.

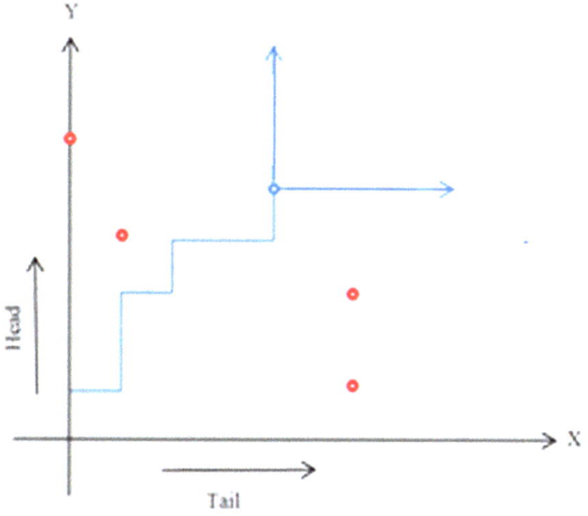

Fig. 2.1 Boundary points

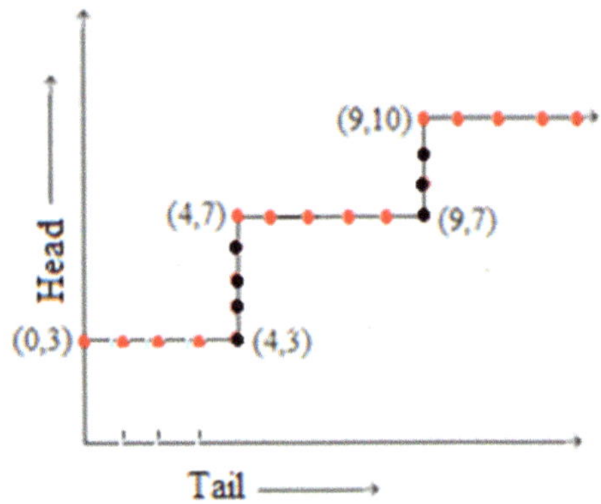

Fig. 2.2 Finite step generalized inverse binomial sampling plan

To get an idea of the accessible and inaccessible points, consider the following plan: stop when 3 heads are observed before observing 4 tails, or 7 heads are observed before observing 9 tails, or 10 heads are observed. This is an example of finite step generalized inverse binomial sampling plan and is indicated in Fig. 2.2.

The lines drawn as per the plan define the absorbing boundary and the red circle points on the absorbing boundary indicate the boundary points. The points falling below the absorbing boundary are the accessible points, and those falling beyond the absorbing boundary are inaccessible points. Besides the red circle points (indicating boundary points), the black circle points are continuation points in the plan.

2.3 Bernoulli Sampling Plans

Consider the experiment of coin tossing with constant success probability 'p'. Suppose the coin is tossed for a given number, say n, of trials and then stopped, irrespective of the results obtained. Let us denote the number of heads observed in the experiment by 'y' and the number of tails observed in the experiment by 'x' so that $x + y = n$. This is referred to as ***binomial (n) sampling plan***. The plan is presented in Fig. 2.3. Vide **ALERT 1** in this context.

In Fig. 2.3, the line $x + y = n$ denotes the absorbing boundary. The red circled points, namely $(0, n), (1, n-1), \ldots, (n, 0)$, on the absorbing boundary are the boundary points. The points that fall below the absorbing boundary are the free points or the accessible points and those falling beyond the absorbing boundary are the inaccessible points, with respect to this particular sampling plan.

Let us now consider a different sampling plan using the coin tossing experiment. Suppose the coin is tossed until 't' heads appear. This plan is referred to as the ***inverse binomial sampling (t) plan***. For this plan, the line $y = t$ is the absorbing boundary, and all the points lying on it are the boundary points. Thus, the plan has a countably infinite number of boundary points. Here, the points falling inside/below

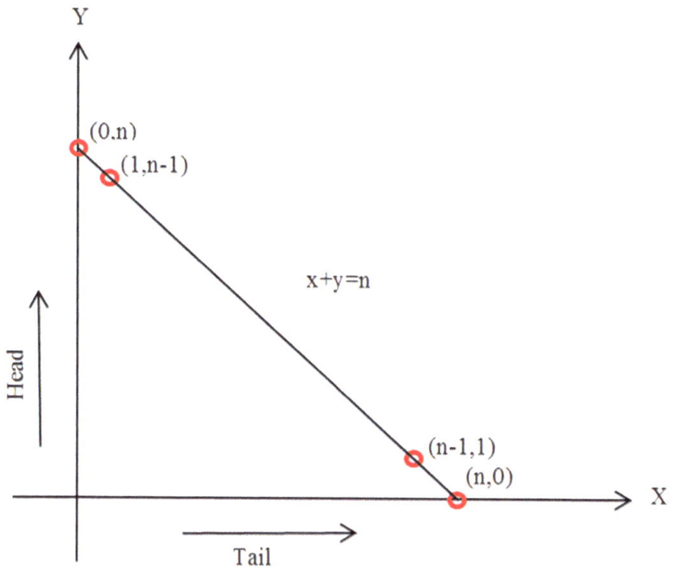

Fig. 2.3 Binomial (n) sampling plan for a specified 'n'

the absorbing boundary are the free points or the accessible points, and those falling beyond/above the absorbing boundary are the inaccessible points.

The plan is presented in Fig. 2.4.

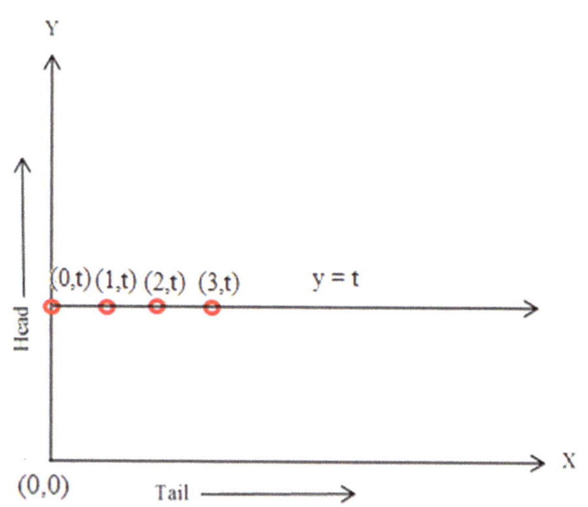

Fig. 2.4 Inverse binomial sampling (t) plan

The two plans discussed above are the traditional plans. Based on these, we can have a family of binomial plans $\mathcal{B} = \{\text{BP}(n): n \geq 1, n \text{ integer}\}$, and a family of inverse binomial plans $\mathcal{IB} = \{\text{IBP}(t): t \geq 1, t \text{ integer}\}$.

2.3.1 *Closure of a Sampling Plan*

Let A denote the collection of all boundary points of a given sampling plan P, and let $\alpha = (x, y) \in A$ be a typical boundary point. Suppose $P(\alpha)$ denotes the probability of reaching α starting from the origin $(0, 0)$. Then, $P(\alpha) = \sum p^y q^x$, where the summation is over all the *accessible* paths from $(0, 0)$ to $\alpha = (x, y)$.

As indicated earlier, $k(\alpha)$ is the number of *accessible* paths from the origin to the point α and we have already noted that trivially $k(\alpha) \leq \binom{x+y}{x}$, for every pair (x, y).

A sampling plan is said to be ***closed*** iff $\sum_{\alpha \in A} P(\alpha) = 1$, for all $p \in (0, 1)$. So, the two plans mentioned above are examples of closed sampling plans. In a closed sampling plan, an experiment terminates with probability 1. For such a plan, the boundary point α and the route that the particle takes to reach α comprise the data for the experiment. A sampling plan which is not closed is said to be '***open***'. However, only closed sampling plans are of interest to an experimenter, and we shall focus on closed plans only.

2.3.2 *Concept of Path Counting*

Recall that $k(\alpha)$ denotes the number of *accessible* paths from the origin to the boundary point α. If all the paths from $(0, 0)$ to α are *free/accessible* paths, then we have $k(\alpha) = \binom{x+y}{x}$. In the case of binomial (n) sampling plan, all paths from $(0, 0)$ to $\alpha = (x, y)$ are free paths when $x + y = n$. But in inverse binomial sampling (t) plan, all paths from $(0, 0)$ to $\alpha = (x, t)$ are *not* free. This could be the case in many situations. In that case, for a boundary point $\alpha = (x, y)$, the number of free paths may be strictly less than $\binom{x+y}{x}$. Intuitively, it would be of interest to count the value of $k(\alpha)$ for a given sampling plan and for a given boundary point α.

The calculation of the value of $k(\alpha)$ may be done by path-counting formula, which is deeply combinatorial in nature. For non-standard sampling plans, a count of $k(\alpha)$ is unavoidable as closure of the plan needs to be verified. An excellent book in this direction is by Mohanty (1979).

For the construction of UMVUEs of parameters of interest, it has been proved that for closed sampling plans arising out of repeated Bernoulli trials, the boundary point α and the path-counting number $k(\alpha)$ jointly define the minimal sufficient statistic.

2.3.3 *Concept of Pull-Down and Push-Up Plans*

In 1-step pull-down plan, we decrease the number of heads at each boundary point of the closed plan by one unit, which means that we pull down each boundary point of the plan by one unit along the Y-axis toward the X-axis. On the other hand, in 1-step push-up plan, we increase the number of heads by one unit, that is, we push up each boundary point of the plan by one unit along the Y-axis away from the X-axis. For example, if we consider Fig. 2.3 of the binomial sampling (n) plan, it would be transformed to Fig. 2.5a after 1-step pull-down, and to Fig. 2.5b after 1-step push-up.

In Fig. 2.5, the solid lines denote the new absorbing boundaries with $x + y = n - 1$ in (a), and $x + y = n + 1$ in (b). The red circle points, namely $(0, n-1), (1, n-2), \ldots, (n-1, 0)$ in (a), and $(0, n+1), (1, n+2), \ldots, (n+1, 0)$ in (b) denote the new boundary points. The new sampling plan in (a) may be denoted by $P_{-1}(\alpha_{-1})$, where $\alpha_{-1} = (x, y-1)$ are the new boundary points, while the new plan in (b) may be denoted by $P_{+1}(\alpha_{+1})$, where $\alpha_{+1} = (x, y+1)$ are the new boundary points. Though it is readily seen that P_{-1} is closed, $P_{+1}(\alpha_{+1})$ is not closed. This is because in the push-up plan, the moving point can reach $(n, 0)$ with probability p^n and thereafter it can move one more step along X-axis with probability p. Thus, under P_{+1}, the particle will escape any of the boundary points and hence not stop with probability 1. We also note that in 1-step pull-down plan, the point $(n, 0)$ sinks below the X-axis.

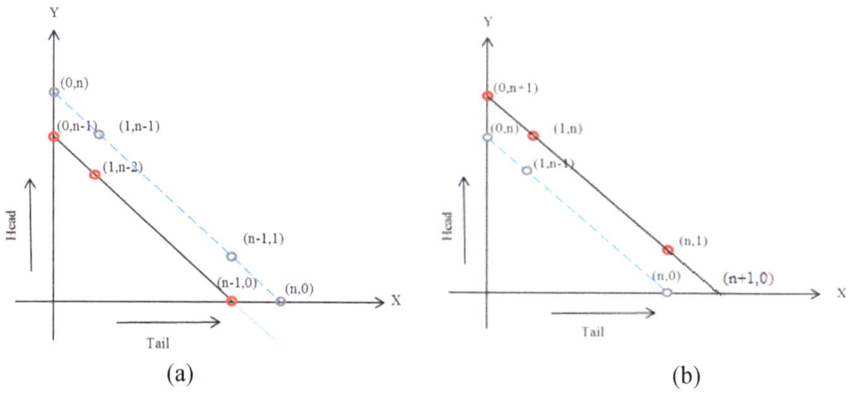

Fig. 2.5 **a** 1-step pull-down binomial sampling (n) plan. **b** 1-step push-up binomial sampling (n) plan

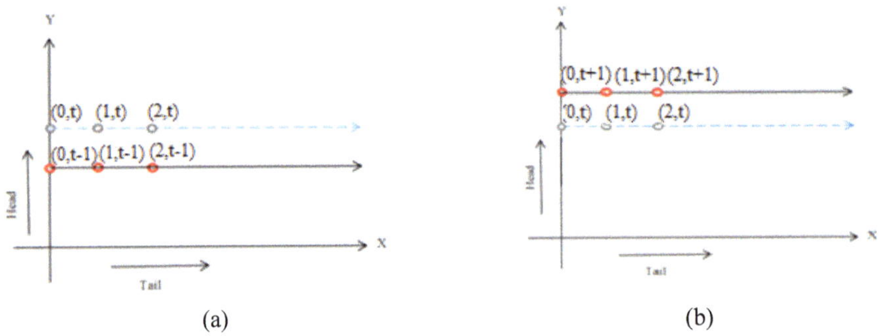

Fig. 2.6 **a** 1-step pull-down inverse binomial sampling (t) plan. **b** 1-step push-up inverse binomial sampling (t) plan

The concept of 1-step pull-down and 1-step push-up plans can be generalized to have pull-down and push-up plans of 2- or more steps, respectively.

In case of the inverse binomial sampling (t) plan, the 1-step pull-down and the 1-step push-up plans transform Figs. 2.4 and 2.6a, b, wherein all points on the line $y = t - 1$ are the new boundary points in 1-step pull-down plan, and all points on the line $y = t + 1$ are the new boundary points in 1-step push-up plan. Here, both P_{-1} and P_{+1} are closed.

Remark 2.3.1 Before we move on to unbiased estimation of 'p', it is revealing to understand how a boundary point can also be a continuation point depending on the path taken. Toward this, we consider a sampling plan P_0 with the following stopping rule:

Stop (i) on getting 4 successive heads; or (ii) on reaching the accessible point (1, 3) by any of the 4 accessible paths, and then getting either a tail (iia) or a head (iib); (iii) on reaching the accessible point (2, 2) [2 heads and 2 tails] followed by 2 heads; or (iv) on reaching the line $y = 3$ without the initial interruptions as described.

In Fig. 2.7, the boundary points of the plan are indicated by red circles. Here, the point (2, 3) plays a dual role in the following sense: Any move from the left side [i.e., (1, 3)] leads to the boundary point (2, 3) and any move from below [i.e., (2, 2)] leads to a continuation point (2, 3). We have highlighted this point by both red and blue colors. The accessible points have been indicated by black circles.

To explain in a different way, the point (2, 3) in the figure is a *boundary point* on the line $y = 3$ along the route: $(0, 0) \ldots \rightarrow (1, 3) \rightarrow (2, 3)$ [Case (iia)], and it is also a *continuation point* along the route: $(0, 0) \ldots \rightarrow (2, 2) \rightarrow (2, 3)$ [Case (iii)].

The analysis of this sampling plan will be taken up in the next section.

Remark 2.3.2 The above sampling plan is seen to resemble an inverse binomial sampling plan ($y = 3$), except for some 'local distortions' in the beginning! We may term such a plan as 'Interrupted IBSP ($y = t$)'.

Fig. 2.7 Sampling plan P_0

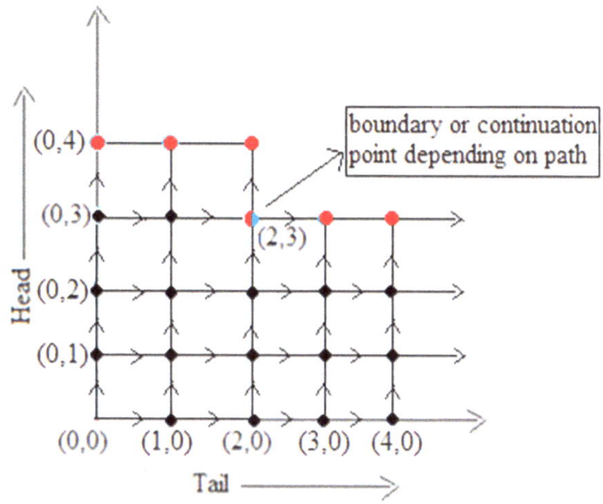

2.4 Unbiased Estimation of Functions of *p*

2.4.1 Unbiased Estimation of p

For any arbitrary closed sampling plan, the pull-down plan can be effectively used for unbiased estimation of p. Recall that the pull-down plan pulls down the whole set of boundary points of the plan by one step along the Y-axis toward the X-axis. Hence, in a pull-down plan, the boundary points *seem* to have come *closer* to the X-axis and, therefore, such a plan is *likely* to be closed as the original plan is so. It is indeed true; however, it still needs a formal proof. The proof is available in a classic paper by Girshick et al. (1946), which introduces much of the concepts and results discussed here.

Since a pull-down sampling plan P_{-1} is closed, we can write

$$\sum_{\alpha_{-1}} P_{-1}(\alpha_{-1}) = 1.$$

Again, $P_{-1}(\alpha_{-1}) = k(x, y-1)p^{y-1}q^x$, so that $\sum_{\alpha_{-1}} k(\alpha_{-1})p^{y-1}q^x = 1$. We can rewrite this as

$$\sum_{\alpha_{-1}} [k(\alpha_{-1})/k(\alpha)]k(\alpha)p^y q^x = p.$$

Since $\sum_{\alpha_{-1}}$ and $\sum_{\alpha}$ are in 1–1 relation, we can replace $\sum_{\alpha_{-1}}$ by $\sum_{\alpha}$ above. This leads to an unbiased estimate of p as $\hat{p} = k(\alpha_{-1})/k(\alpha)$; that is, an unbiased estimate of 'p' is given by the ratio of the two path counts.

For example, in the case of binomial sampling (n) plan, where $x + y = n$, we have $k(\alpha) = \binom{n}{x}$ and $k(\alpha_{-1}) = \binom{n-1}{x}$, so that $\hat{p} = y/n$, which is the usual estimator of p. Note that here 'y' denotes the count of heads in the plan [Vide **ALERT 1**].

Again, in case of inverse binomial sampling (t) plan, when we apply the 1-step pull-down approach, we get $k(\alpha) = \binom{t+x-1}{x}$ and $k(\alpha_{-1}) = \binom{t+x-2}{x}$, which gives $\hat{p} = (t-1)/(x+t-1)$. Recall that, x gives the tail count in the plan and this formula works only when $t > 1$. In other words, there does not exist any unbiased estimator for p when $t = 1$.

2.4.2 *Unbiased Estimation of p^{-1}*

Let P_0 be an arbitrarily closed sampling plan. After 1-step push-up, the new plan is denoted by P_{+1} with α_{+1} as a typical boundary point (arising out of the boundary point $\alpha = (x, y)$ of P_0). This α_{+1} naturally has the representation $(x, y+1)$, and $k(\alpha_{+1})$ is the number of accessible paths from $(0, 0)$ to α_{+1}, with reference to P_{+1}. Suppose the original plan P_0 is so chosen that the 1-step push-up plan P_{+1} is also closed. Then $\frac{1}{p}$ can be unbiasedly estimated by $k(\alpha_{+1})/k(\alpha)$, where α, as is understood from above, is a typical boundary point of P_0. The proof of this observation is indicated below:

As P_{+1} is assumed to be closed,

$$\sum_{\alpha_{+1}} k(\alpha_{+1}) p^{y+1} q^{x} = 1$$

$$\Rightarrow \sum_{\alpha_{+1}} \left[k(\alpha_{+1})/k(\alpha)\right] k(\alpha) p^{y} q^{x} = \frac{1}{p}$$

$$\Rightarrow \widehat{\left(\frac{1}{p}\right)} = k(\alpha_{+1})/k(\alpha).$$

Bose and Sinha (1985) proved that the closure of a push-up plan P_{+1} is necessary as well for unbiased estimation of $1/p$.

Remark 2.4.1 It is interesting to note that once a 1-step push-up plan is seen to be closed, it will remain so for all subsequent higher level push-ups! Consequently, by using the 2-step push-up plan, we can find unbiased estimate of $\frac{1}{p^2}$ as $k(\alpha_{+2})/k(\alpha)$. Proceeding this way, we can unbiasedly estimate any negative integer power of p. Note that unbiased estimation of $1/p^2$ is necessary for estimation of variance of the estimate of $1/p$.

Remark 2.4.2 Binomial sampling plan (n) provides unbiased estimators of p, p^2, p^3, etc., up to, and including, p^n, and no negative integer power of p is amenable to such unbiased estimation. On the other hand, inverse binomial sampling (t) plan provides unbiased estimators of p, p^2, etc., up to p^{t-1} and, at the same time, it provides unbiased estimators for ***all*** negative integer powers of p! Later we will see many more appealing features of this latter type of plan in the context of unbiased estimation of functions of the parameter p.

Let us now recall the sampling plan P_0 given by Fig. 2.7. The boundary points of the plan, their path counts, and the associated probabilities are given in Table 2.1.

It must be noted that to reach (3, 3) there are two distinct paths, namely

(i) reaching (2, 2) [6 free paths] and going to (2, 3) and then to (3, 3);
(ii) reaching (3, 2) [10 free paths] and then going to (3, 3).

The total associated probability is $p^4 + 4p^4q + 4p^3q^2 + 6p^4q^2 + 16p^3q^3 + \sum_{x=4}^{\infty} a(x)p^3q^x$, which simplifies to 1. Hence, the plan is closed.

For unbiased estimation of 'p', consider the 1-step pull-down plan P_{-1} represented in Fig. 2.8.

The boundary points, their path counts, and the corresponding probabilities are shown in Table 2.2.

Table 2.1 Boundary points, path counts, and associated probabilities in sampling plan P_0

Boundary point (α_0)	Path count $k(\alpha_0)$	Probability $P(\alpha_0)$
(0, 4)	1	p^4
(1, 4)	4	$4\,p^4q$
(2, 3)	4	$4\,p^3q^2$
(2, 4)	6	$6\,p^4q^2$
(3, 3)	16	$16\,p^3q^3$
$(x, 3); x \geq 4$	$a(x) = (x+1)(x+2)/2$	$a(x)\,p^3q^x; x \geq 4$

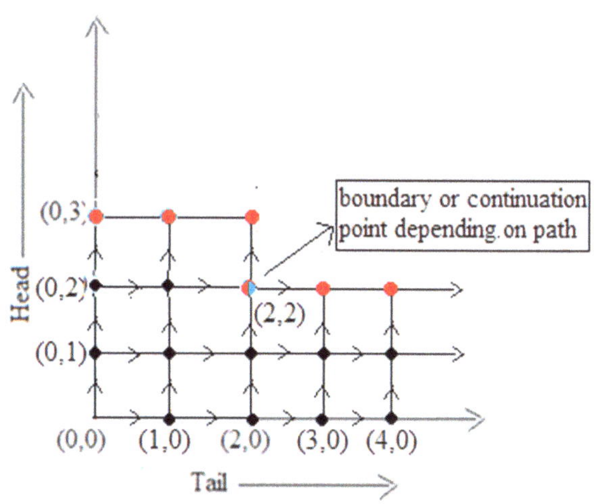

Fig. 2.8 1-step pull-down plan P_{-1}

Table 2.2 Boundary points, path counts, and associated probabilities in sampling plan P_{-1}

Boundary point (α_{-1})	Path count $k(\alpha_{-1})$	Probability $P(\alpha_{-1})$
(0, 3)	1	p^3
(1, 3)	3	$3\,p^3q$
(2, 2)	3	$3\,p^2q^2$
(2, 3)	3	$3\,p^3q^2$
(3, 2)	7	$7\,p^2q^3$
$(x, 2); x \geq 4$	$a^*(x) = (x + 1)$	$a^*(x)\,p^2q^x; x \geq 4$

Table 2.3 Estimates $\hat{p}$ of p

Boundary point	$\hat{p}$
(0, 4)	0
(1, 4)	3/4
(2, 3)	3/4
(2, 4)	1/2
(3, 3)	7/16
$(x, 3); x \geq 4$	$2/(x + 2)$

To reach (3, 2), there are two disjoint directed paths, namely

(i) reaching (2, 1) [3 free paths] and going to (2, 2) from there and then to (3, 2);
(ii) reaching (3, 1) [4 free paths] and then going to (3, 2) from there.

The total probability to culminate at a boundary point is $p^3 + 3p^3q + 3p^2q^2 + 3p^3q^2 + 7p^2q^3 + \sum_{x=4}^{\infty} a(x)p^2q^x$, which simplifies to 1. Hence the plan is closed.

By the path-counting method, the estimates of p at the boundary points of P_0 are given in Table 2.3.

Remark 2.4.3 It may be noted that the Interrupted IBSP ($y = 3$) is also closed under push-up, and hence we can unbiasedly estimate $1/p$ as well. We omit the details.

2.4.3 *Unbiased Estimation of p, q, and pq under Two Non-traditional Sampling Plans*

Here we illustrate two non-traditional sampling plans for the estimation of p, q, and pq.

Plan 1: *Perform a coin tossing experiment till 3 heads or 5 tails appear.*

The boundary points of this plan are shown in Fig. 2.9.

The red circled points are the boundary points of the plan, and the point (5, 3) is an inaccessible point. The closure property of Plan 1 and the corresponding 1-step pull-down sampling plan P_{-1} can be easily verified.

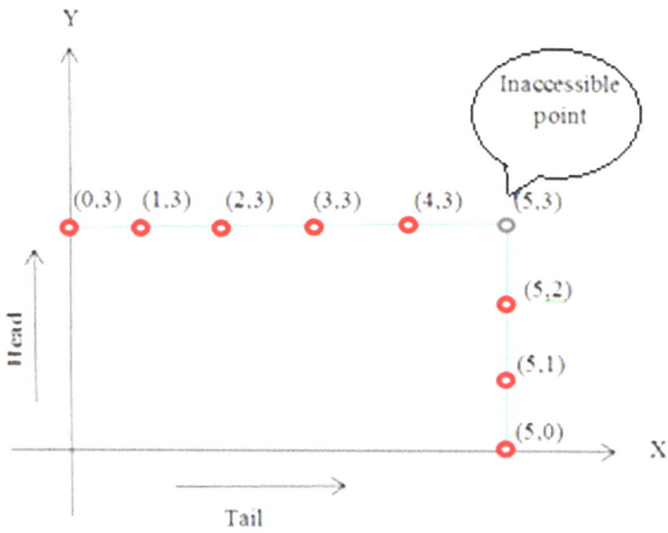

Fig. 2.9 Boundary points for Plan 1

Using the plan P_{-1}, the estimate $\hat{p}$ of p is given in Table 2.4. Note that it is a proper estimator.

Likewise, carrying out the 1-step pull-down plan along the X-axis, toward the Y-axis, we can find unbiased estimate of q. For estimation of pq, we use the *doubly pulled-down* plan $P_{-1,-1}$, wherein we pull down one step along Y-axis and one step along X-axis. Let the new boundary points be α^*_{-1} (Table 2.5).

Remark 2.4.4 Note that pq has the natural bounds $0 \leq pq \leq 1/4$. However, in the unbiased estimates computed above, except for the boundary points (5, 0), (5, 1), and (0, 3), all other boundary points yield estimates greater than 1/4. This happens whatever be the nature of the closed sampling plan we start with—there will be at least one boundary point for which the estimate will exceed 1/4 (vide Tarafdar, 2016).

Table 2.4 Estimate $\hat{p}$ of p

α	α_{-1}	$k(\alpha)$	$k(\alpha_{-1})$	$\hat{p} = k(\alpha_{-1})/k(\alpha)$
(5, 0)	–	1	0	0
(5, 1)	(5, 0)	5	1	1/5
(5, 2)	(5, 1)	15	5	1/3
(4, 3)	(5, 2)	15	15	1
(3, 3)	(3, 2)	10	6	3/5
(2, 3)	(2, 2)	6	3	1/2
(1, 3)	(1, 2)	3	2	2/3
(0, 3)	(0, 2)	1	1	1

Table 2.5 Estimate $\widehat{pq}$ of pq

α	α^*_{-1}	$k(\alpha)$	$k(\alpha^*_{-1})$	$\widehat{pq} = k(\alpha^*_{-1})/k(\alpha)$
(5, 0)	–	1	0	0
(5, 1)	(4, 0)	5	1	1/5
(5, 2)	(4, 1)	15	4	4/15
(4, 3)	(3, 2)	15	6	2/5
(3, 3)	(2, 2)	10	3	3/10
(2, 3)	(1, 2)	6	2	1/3
(1, 3)	(0, 3)	3	1	1/3
(0, 3)	–	1	0	0

In the above, we have overlooked this phenomenon. We will continue to work on a variety of estimation problems by considering suitable binomial sampling plans, completely ignoring this aspect.

It may be noted that in the estimation of positive or negative integer powers of p (or of q), as done under Plan 2 below, the above problem does not arise. There the unbiased estimation by the path-counting formula yields desirable results in all cases.

Plan 2: *Perform a coin tossing experiment till 3 heads and 5 tails appear.*

In this case, the sampling plan can be represented diagrammatically as shown in Fig. 2.10.

The boundary points of the plan are $(5, 3)$, $\{(x, 3) \text{ for } x \geq 6\}$, and $\{(5, y) \text{ for } y \geq 4\}$. For $\alpha = (5, 3)$, all the paths from the point (0, 0) are free paths. Hence $k(\alpha)$ will be $\binom{5+3}{5} = 56$. The boundary points of the form $\{(x, 3),\ x \geq 6\}$ cannot be reached from the left-hand side. These can be reached

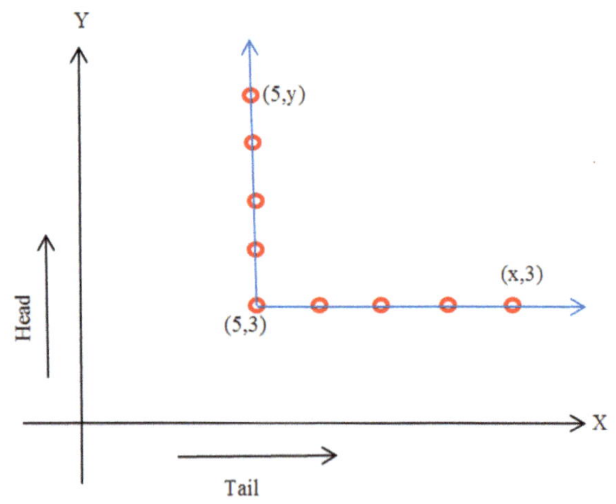

Fig. 2.10 Boundary points for Plan 2

only through movement in the upward direction. For example, if we consider $\alpha = (x, 3)$, this can be reached only through the point $(x, 2)$. Hence, in this case, the value of $k(\alpha)$ is $\binom{x+2}{2}$. Similarly, boundary points of the type $\{(5, y), y \geq 4\}$, can be reached only from the left-hand side. Hence, for this type of boundary points, the value of $k(\alpha)$ is $\binom{4+y}{4}$.

To show closure of the plan, we may argue in the following way:

(i) The boundary points of the inverse Bernoulli plan described by '$y = 3$' include all the boundary points $\{(x, 3); x \geq 5\}$, and the path counts remain the same.
(ii) The boundary points of the inverse Bernoulli plan described by '$x = 5$' include all the boundary points $\{(5, y); y \geq 3\}$, and the path counts remain the same.
(iii) The two plans in (i) and (ii) together comprise Plan 2, given by Fig. 2.10, and Plan 1, given by Fig. 2.9.

Since Fig. 2.9 refers to a closed plan, we necessarily have closure of the Plan 2.

We now focus on estimation of 'p'. We carry out 1-step pull-down plan along the Y-axis, toward the X-axis, and, compute the values of $k(\alpha_{-1})$ or the modified plan P_{-1}. The closure property of P_{-1} is readily verified.

Table 2.6 gives the estimate $\hat{p}$ of p for different boundary points α.

To estimate pq, one approach would be to estimate p and p^2 first using 1-step pull-down and 2-step pull-down sampling plans toward the X-axis, respectively. Another way would be to estimate it directly by doubly pulled-down plan using 1-step pull-down sampling plan toward X-axis and that toward Y-axis simultaneously. Denoting by $\alpha^*_{-1} = (x - 1, y - 1)$ a typical boundary point of the modified plan, we compute $k(\alpha^*_{-1})$ for all the boundary points to estimate pq. Here we skip the details.

Remark 2.4.5 This kind of coin tossing plans have been introduced and studied by Sinha and Sinha (1975). In fact, a few other such binomial sampling plans are also introduced by them.

Table 2.6 Estimation of p

α	α_{-1}	$k(\alpha)$	$k(\alpha_{-1})$	$\hat{p} = k(\alpha_{-1})/k(\alpha)$
(5, 3)	(5, 2)	56	21	0.375
$(x, 3); x \geq 6$	$(x, 2); x \geq 6$	$\binom{x+2}{2}$	$\binom{x+1}{1}$	$\frac{2}{x+2}$
$(5, y); y \geq 4$	$(5, y); y \geq 3$	$\binom{4+y}{y}$	$\binom{3+y}{y-1}$	$\frac{y}{4+y}$

2.4.4 Unbiased Estimation of $p^a q^b$

We start by stating an interesting result toward estimation of $p^a q^b$ based on a closed sampling plan. (Vide the **ALERT 1** regarding positions of a and b.)

Result 2.4.1: A necessary and sufficient condition for the existence of a sampling plan to unbiasedly estimate $p^a q^b$ where a and b are finite positive integers, is that there exists at least one point on the line $x + y = a + b$ which is an accessible/boundary point.

While we do not venture to provide a proof of this result, we present a non-trivial example of a plan and a choice of (a, b) to get to the point.

Consider the plan with boundary points (indicated by red circles) shown in Fig. 2.11, and the choice $(a = 2, b = 3)$.

Note that on the line $x + y = 5$, the point (2, 3) is inaccessible and so are all other points except (4, 1) and (5, 0), which are boundary points. Also there are 4 points on the line $x + y = 4$ which are all boundary points.

We express p^2q^3 as a function of the boundary points available in the plan.

It follows trivially that $p^2q^3 = pq^3(1 - q) = pq^3 - pq^4$ and the corresponding points (3, 1) and (4, 1) are both boundary points. Hence unbiased estimation is possible by using corresponding indicator function for each term.

From the result, it is apparent that the application of appropriate pull-down plans will serve to provide an unbiased estimate of $p^a q^b$, for $a > 0$ and $b > 0$. It must be noted that we are ignoring the fact that such estimates may suffer from the property of being 'improper', irrespective of the nature of the closed sampling plan we start with.

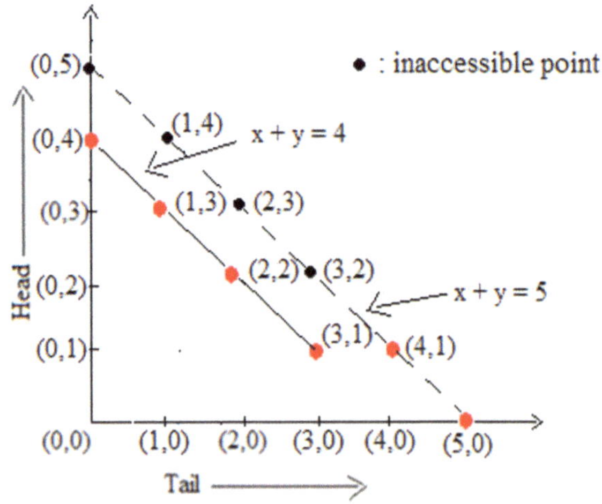

Fig. 2.11 Boundary points of the plan

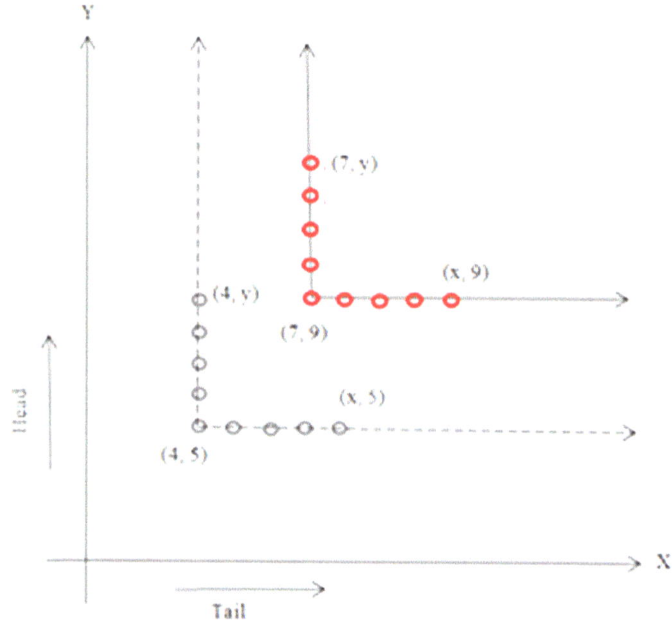

Fig. 2.12 Showing the boundary points (as black circle points) of the sampling plan $P_{(0,0)}$, and the boundary points (as red circle points) of the modified plan $P_{(+3,+4)}$

Let us now focus on the problem of estimating $p^a q^b$, when a or b or both are finite negative integers. Consider the following combinations of (a, b): $(4, -3)$, $(-4, 3)$, $(-4, -3)$. We consider the sampling plan $P_{(0,0)}$, where a coin tossing experiment is performed till 5 heads and 4 tails are obtained. The boundary points of the sampling plan are shown by black circle points in Fig. 2.12.

At this stage, it may be noted that Gupta (1967) established that a necessary condition for unbiased estimation of a negative power of 'p' is that the initial plan must be unbounded in X-direction. Likewise, it follows that the plan must be unbounded in Y-direction in the context of unbiased estimation of any negative power of 'q'.

Therefore, the plan in Fig. 2.12 qualifies, to start with, as a possible plan to provide unbiased estimates of all the three parametric functions simultaneously.

2.4.4.1 Unbiased Estimation of p^4q^{-3}

On binomial expansion of $p^4 = (1-q)^4$, we have

$$p^4q^{-3} = q^{-3} - 4q^{-2} + 6q^{-1} - 4 + q.$$

Now, using 1-step pull-down plan along X-axis toward the Y-axis, we can estimate q; using 1-step push-up plan away from Y-axis, we estimate q^{-1}; using 2-step push-up plan away from Y-axis we can estimate q^{-2}; finally, using 3-step push-up plan away from Y-axis, we can estimate q^{-3}. Substituting these estimates in the expression for p^4q^{-3}, we get an unbiased estimate of the same.

It should be noted that all the relevant 'derived' plans are closed under the plan.

2.4.4.2 Unbiased Estimation of $p^{-4}q^3$

In this case, we expand $q^3 = (1 - p)^3$ to get

$$p^{-4}q^3 = p^{-4} - 3p^{-3} + 3p^{-2} - p^{-1}.$$

Then, through arguments analogous to that in the previous subsection, we get an unbiased estimate of $p^{-4}q^3$.

2.4.4.3 Estimation of $p^{-4}q^{-3}$

Unlike in the two previous subsections, here we cannot get a finite expansion of $p^{-4}q^{-3}$. This makes estimation of $p^{-4}q^{-3}$ comparatively difficult.

We use 4-step push-up sampling plan along Y-axis and 3-step push-up sampling plan along X-axis. The boundary points of the modified plan $P_{(+3,+4)}$ are indicated by red circle points in Fig. 2.12.

The boundary points of $P_{(0,0)}$ are (4, 5), $\{(4, y), y \geq 6\}$ and $\{(x, 5), x \geq 5\}$, while the boundary points of $P_{(+3,+4)}$ are (7, 9), $\{(7, y), y \geq 10\}$ and $\{(x, 9), x \geq 8\}$. At each boundary point α of $P_{(0,0)}$, the estimate of $\theta = p^{-4}q^{-3}$ is given by $\hat{\theta} = k(\alpha^*_{+4,+3})/k(\alpha)$, where $\alpha^*_{+4,+3}$ denotes the corresponding boundary point of $P_{+3,+4}$ (Table 2.7).

The closure property of the plans can be easily verified.

Remark 2.4.6 We must note that our starting sampling plan indeed qualifies to provide closure of all these 'derived' plans at the same time. Many other closed sampling plans do admit of these closure properties as well. However, an inverse binomial sampling plan of the type $y = 4$, for example, will not be suitable for this

Table 2.7 Estimation of $\theta = p^{-4}q^{-3}$

α	$k(\alpha)$	$\alpha^*_{+4,+3}$	$k(\alpha^*_{+4,+3})$	$\hat{\theta}$
(4, 5)	126	(7, 9)	6435	51.07
$(x, 5); x \geq 5$	$\binom{x+4}{4}$	$(x, 9); x \geq 7$	$\binom{x+8}{8}$	$\binom{x+8}{8}/\binom{x+4}{4}$
$(4, y); y \geq 6$	$\binom{3+y}{3}$	$(7, y); y \geq 10$	$\binom{6+y}{6}$	$\binom{6+y}{6}/\binom{3+y}{3}$

purpose. This follows from Gupta (1967), who showed that a necessary condition for unbiased estimation of a negative power of 'p' is that the initial plan must be unbounded in X-direction. Similarly, the plan must be unbounded in Y-direction for unbiased estimation of any negative power of 'q'.

2.5 Unbiased Estimation of Parameters in a Trinomial Set-Up

Having exposed the readers to the mysterious world of repeated Bernoulli trials, we now move on to parameter estimation in a trinomial set-up. For instances of occurrences of trinomial distributions, as well as tetranomial distributions (to be dealt with in Sect. 2.6), in real-life situations, we refer the readers to Rao (1952). Rao used such distributions in the classification of individuals with respect to phenotypes of blood groups.

To start with, we describe the notations and terminologies below. The readers are now advised to recall the usual notations generally followed in explaining binomial and trinomial distributions. In contrast to **ALERT 1** it is advisable that we refer to:

ALERT 2: p, q, and r respectively stand for the probabilities associated with a random outcome which is represented along the X-direction, or Y-direction, or Z-direction.

From now on we will be guided by these specifications.

Consider $(V_i, i = 1, 2, \ldots)$ to be a sequence of trials, each of which has three possible outcomes with corresponding probabilities p, q, r, such that $0 < p, q, r < p + q + r = 1$. Then, any experiment can be represented as a lattice path in the (X, Y, Z)-space, where a particle moves from the origin one step forward along X-axis if an observation is an outcome having probability p, one step forward along Y-axis if it is an outcome having probability q and one step forward along Z-axis if it is an outcome having probability r. A stopping rule can be looked upon as a sequence of functions φ_k, where φ_k is a function of $(V_1, \ldots, V_k)$. Given $(v_1, \ldots, v_k)$, each φ_k takes the value 0 or 1: $\varphi_k(v_1, \ldots, v_k) = 1$ indicates that one more observation should be taken, and $\varphi_k(v_1, \ldots, v_k) = 0$ indicates that we stop at that stage. As in case of Bernoulli trials, a point $\alpha = (x, y, z)$ denotes a continuation point if there exists one sequence of realization $\left(v_1, \ldots, v_{x+y+z}\right)$ leading to α such that $\varphi_j\left(v_1, \ldots, v_j\right) = 1\ \forall j \le x+y+z$. A point $\alpha = (x, y, z)$ is a boundary point if there exists one sequence of realization $\left(v_1, \ldots, v_{x+y+z}\right)$ leading to α such that $\varphi_j\left(v_1, \ldots, v_j\right) = 1$ for all $j < x + y + z$ and $\varphi_{x+y+z}\left(v_1, \ldots, v_{x+y+z}\right) = 0$. A point may be a boundary point or a continuation point depending on the path. A point is an accessible point if it is either a boundary point or a continuation point. Points which are not accessible are inaccessible points.

2.5.1 *Parameter Estimation in a Trinomial Distribution*

Consider an experiment which has three possible outcomes, say O_1, O_2, and O_3, with probabilities p, q, and r, respectively, such that $0 < p, q, r < p + q + r = 1$. Suppose, we repeat the experiment n (finite) times. Let x denote the number of times O_1 occurs, y the number of times O_2 occurs, and z the number of times O_3 occurs, so that $x + y + z = n$. A sampling plan involving such an experiment may be denoted by $P_{(0,0,0)}$. This notation is supposed to provide a convenient way of describing other sampling plans in similar context. For a fixed value of x, say $x = t$ $(0 \leq x \leq n)$, the boundary points of the plan are points on the line $y+z = n-t$. The boundary points are graphically represented in Fig. 2.13. In the diagram, corresponding to a fixed value t of x, the boundary points are of the form (t, y, z), where y and z are integers satisfying $0 \leq y, z \leq n - t$ such that $y + z = n - t$. Some of the boundary points are denoted by red circular points. For $x = n$, there is only one boundary point, namely $(n, 0, 0)$.

For a typical boundary point $\alpha = (x, y, z)$ of the plan, we have

$$k(\alpha) = \frac{n!}{x!y!z!}.$$

It is, then, readily seen that the plan is closed as $\sum_{\alpha} k(\alpha)p^x q^y r^z = 1$.

We shall henceforth denote a boundary point $\alpha = (x, y, z)$ of the plan by $\alpha_{(0,0,0)}$.

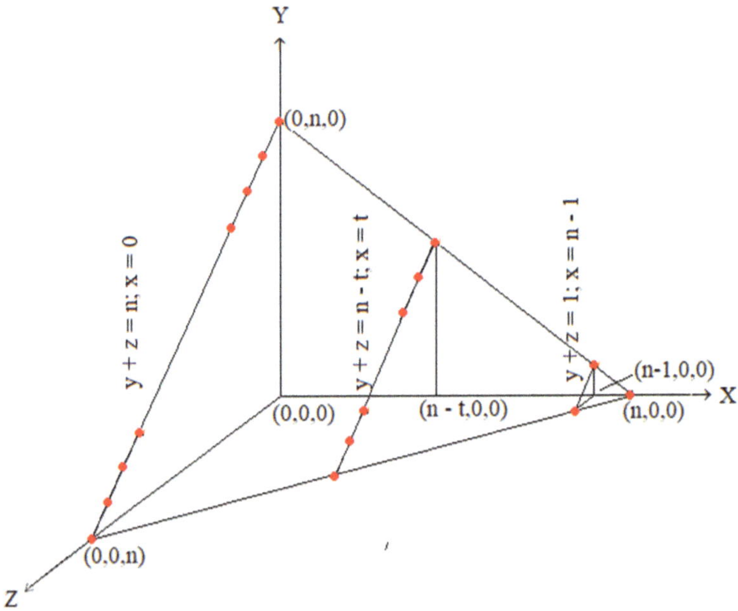

Fig. 2.13 Trinomial sampling plan

To estimate p, we consider 1-step pull-down plan along the X-axis toward the (Y, Z)-plane. Let us denote the new boundary points as $\alpha_{(-1,0,0)} = (x-1, y, z)$, and the plan as $P_{(-1,0,0)}$. It is easily seen that the plan is closed.

To get an unbiased estimator of p, we argue in the following way:

$$\sum_{P_{(0,0,0)}} k\left(\alpha_{(0,0,0)}\right)p^x q^y r^z = 1 \quad \text{since } P_{(0,0,0)} \text{ is closed}$$

$$\Rightarrow \sum_{P_{(-1,0,0)}} k\left(\alpha_{(-1,0,0)}\right)p^{x-1} q^y r^z = 1, \quad \text{since } P_{(-1,0,0)} \text{ is closed}$$

$$\Rightarrow \sum_{P_{(-1,0,0)}} k\left(\alpha_{(-1,0,0)}\right)p^x q^y r^z = p$$

$$\Rightarrow \sum_{P_{(-1,0,0)}} \frac{k\left(\alpha_{(-1,0,0)}\right)}{k\left(\alpha_{(0,0,0)}\right)} k\left(\alpha_{(0,0,0)}\right)p^x q^y r^z = p.$$

Hence, $\hat{p} = \frac{k\left(\alpha_{(-1,0,0)}\right)}{k\left(\alpha_{(0,0,0)}\right)} = \frac{\frac{(n-1)!}{(x-1)!y!z!}}{\frac{n!}{x!y!z!}} = \frac{x}{n}$ is an unbiased estimator of p.

However, unbiased estimator of $1/p$ cannot be obtained as the 1-step push-up of trinomial sampling plan is not closed.

For unbiased estimation of q and r, we use 1-step pull-down plan $P_{(0,-1,0)}$ along the Y-axis toward the (X, Z)-plane and 1-step pull-down plan $P_{(0,0,-1)}$ along the Z-axis toward the (X, Y)-plane, respectively. Then, the unbiased estimators of q and r are obtained as

$$\hat{q} = \frac{k\left(\alpha_{(0,-1,0)}\right)}{k\left(\alpha_{(0,0,0)}\right)} = \frac{\frac{(n-1)!}{x!(y-1)!z!}}{\frac{n!}{x!y!z!}} = \frac{y}{n}$$

$$\hat{r} = \frac{k\left(\alpha_{(0,0,-1)}\right)}{k\left(\alpha_{(0,0,0)}\right)} = \frac{\frac{(n-1)!}{x!y!(z-1)!}}{\frac{n!}{x!y!z!}} = \frac{z}{n}.$$

Now, consider estimation of pq. For this, we need a doubly pull-down sampling plan $P_{(-1,-1,0)}$ whereby we pull down one step along the X-axis and one step along the Y-axis simultaneously toward the Z-axis. This gives the unbiased estimator of pq as

$$\widehat{pq} = \frac{k\left(\alpha_{(-1,-1,0)}\right)}{k\left(\alpha_{(0,0,0)}\right)} = \frac{\frac{(n-2)!}{(x-1)!(y-1)!z!}}{\frac{n!}{x!y!z!}} = \frac{xy}{n(n-1)}.$$

Similarly, to estimate pqr, we use 1-step pull-down sampling plan along all the three directions, which results in the plan $P_{(-1,-1,-1)}$. Hence, the unbiased estimator is

$$\widehat{pqr} = \frac{k\left(\alpha_{(-1,-1,-1)}\right)}{k\left(\alpha_{(0,0,0)}\right)} = \frac{\frac{(n-3)!}{(x-1)!(y-1)!(z-1)!}}{\frac{n!}{x!y!z!}} .$$
$$= \frac{xyz}{n(n-1)(n-2)}$$

2.5.2 Parameter Estimation in an Inverse Trinomial Distribution

In the inverse trinomial sampling plan, say $P_{(t,0,0)}$, we repeat the experiment with three possible outcomes, O_1, O_2, and O_3, with probabilities p, q, and r, respectively ($0 < p, q, r < p+q+r = 1$), till, for example, the outcome O_1 occurs t times, where t is a fixed/specified positive integer. Let us denote this plan by $P_{(t,0,0)}$. If $\alpha_{(0,0,0)} = (x, y, z)$ denotes a typical boundary point of the plan, we have

$$k\left(\alpha_{(0,0,0)}\right) = \frac{(t+y+z-1)!}{(t-1)!y!z!}.$$

In this case, the (Y, Z)-plane serves as the collection of countably infinite number of boundary points. The plan is shown in Fig. 2.14. The red circle points denote the boundary points, and each square in the (Y, Z)-plane is of unit length.

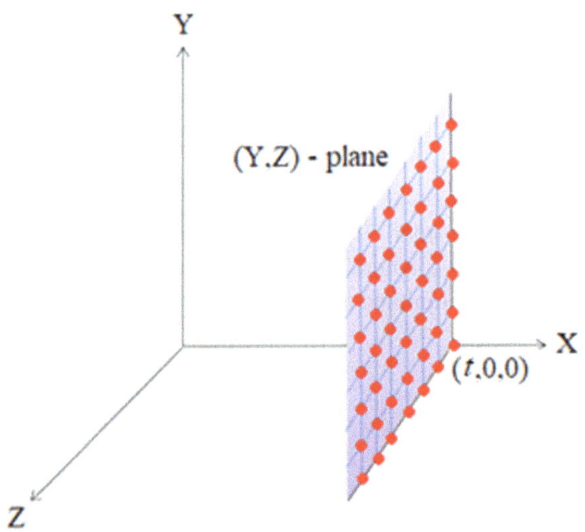

Fig. 2.14 Inverse trinomial plan $x = t$

We note that the plan $P_{(t,0,0)}$ is closed. This ensures that the 1-step pulled-down plan $P_{(t-1,0,0)}$ along the X-axis toward the (Y, Z)-plane is also closed since the latter is a plan with boundary points on the (Y, Z)-plane corresponding to $x = t - 1$. Let $\alpha_{(-1,0,0)}$ denote a typical boundary point of $P_{(t-1,0,0)}$.

Using the routine path-counting method, we obtain an unbiased estimator of p as

$$\hat{p} = \frac{k\left(\alpha_{(-1,0,0)}\right)}{k\left(\alpha_{(0,0,0)}\right)} = \frac{\frac{(t+y+z-2)!}{(t-2)!y!z!}}{\frac{(t+y+z-1)!}{(t-1)!y!z!}} = \frac{t-1}{t+y+z-1}.$$

To estimate $1/p$, we consider the 1-step push-up plan along the X-axis away from the (Y, Z)-plane, denoted by $P_{(t,0,0)}$. It has boundary points on the (Y, Z)-plane corresponding to $x = t + 1$. Hence, the plan is closed, and therefore $1/p$ is unbiasedly estimable. Denoting the new boundary points as $\alpha_{(+1,0,0)} = (t + 1, y, z)$, the unbiased estimator of $1/p$ is given by

$$\widehat{1/p} = \frac{k\left(\alpha_{(+1,0,0)}\right)}{k\left(\alpha_{(0,0,0)}\right)} = \frac{\frac{(t+y+z)!}{t!y!z!}}{\frac{(t+y+z-1)!}{(t-1)!y!z!}} = \frac{t+y+z}{t}.$$

Unbiased estimation of q and r from the inverse trinomial sampling plan $P_{(t,0,0)}$ using 1-step pull-down plans is feasible as the plans are closed. So, using 1-step pull-down plan along the Y-axis and along the Z-axis, respectively, we get

$$\hat{q} = \frac{y}{t+y+z-1} \text{ and } \hat{r} = \frac{z}{t+y+z-1}.$$

However, estimation of $1/q$ (or $1/r$) from the inverse trinomial sampling plan $P_{(t,0,0)}$ using the 1-step push-up plan $P_{(t,+1,0)}$ (or $P_{(t,0,+1)}$), is not feasible as such plans are not closed.

To estimate pq, we consider doubly pull-down sampling plan $P_{(t-1,-1,0)}$, that is, 1-step pull-down along the X-axis and that along the Y-axis simultaneously, both toward the Z-axis. Denoting a new boundary point as $\alpha_{(-1,-1,0)}$, we get the unbiased estimator as

$$\widehat{pq} = \frac{k\left(\alpha_{(-1,-1,0)}\right)}{k\left(\alpha_{(0,0,0)}\right)} = \frac{\frac{(t+y+z-3)!}{(t-2)!(y-1)!z!}}{\frac{(t+y+z-1)!}{(t-1)!y!z!}}$$
$$= \frac{y(t-1)}{(t+y+z-1)(t+y+z-2)}.$$

Similarly, to estimate pqr, we take 1-step pull-down sampling plan along X-axis, Y-axis, and Z-axis simultaneously which results in the plan $P_{(t-1,-1,-1)}$.

We denote the new boundary point as $\alpha_{(-1,-1,-1)}$. Then, the unbiased estimator of pqr is obtained as

$$\widehat{pqr} = \frac{k\left(\alpha_{(-1,-1,-1)}\right)}{k\left(\alpha_{(0,0,0)}\right)} = \frac{\frac{(t+y+z-4)!}{(t-2)!(y-1)!(z-1)!}}{\frac{(t+y+z-1)!}{(t-1)!y!z!}} \\ = \frac{yz(t-1)}{(t+y+z-1)(t+y+z-2)(t+y+z-3)}.$$

Remark 2.5.1 We may estimate p, pq, pqr in the trinomial distribution using the first principle method, too. This is for verification by the beginners. However, the path-counting method allows us to get unbiased estimators of other complicated forms of the parameters. This unified method of unbiased estimation in terms of ratios of path counts is worth learning and preaching!!!

2.5.3 *Unbiased Estimation of $p^a q^b r^c$*

Consider the trinomial sampling plan $P_{(0,0,0)}$ discussed in Sect. 2.5.1. Suppose we want to estimate $p^a q^b r^c$, where a, b, and c are integers.

First, let us consider a, b, and c to be positive integers satisfying $a + b + c \leq n$. To unbiasedly estimate $\theta = p^a q^b r^c$, we consider the pull-down sampling plan which simultaneously pulls down a steps along the X-axis, b steps along the Y-axis, and c steps along the Z-axis. The resulting plan is $P_{(-a,-b,-c)}$. Then, by the path-counting method, the unbiased estimator of θ is

$$\hat{\theta} = \frac{k\left(\alpha_{(-a,-b,-c)}\right)}{k\left(\alpha_{(0,0,0)}\right)},$$

where $\alpha_{(0,0,0)} = (x, y, z)$ and $\alpha_{(-a,-b,-c)} = (x - a, y - b, y - c)$ represent the boundary points of the original plan $P_{(0,0,0)}$ and the pulled-down plan $P_{(-a,-b,-c)}$, respectively.

If $P_{(0,0,0)}$ be a trinomial sampling plan with x, y, and z satisfying $x + y + z = n$, the estimator $\hat{\theta}$ takes the form

$$\hat{\theta} = \frac{\frac{(n-a-b-c)!}{(x-a)!(y-b)!(z-c)!}}{\frac{n!}{x!y!z!}}.$$

On the other hand, if $P_{(t,0,0)}$ be the inverse trinomial sampling plan $x = t$, the estimator is

$$\hat{\theta} = \frac{\frac{(t+y+z-1-a-b-c)!}{(t-1-a)!(y-b)!(z-c)!}}{\frac{(t+y+z-1)!}{(t-1)!y!z!}}.$$

Remark 2.5.2 It is noteworthy that in the trinomial sampling plan, for the existence of the estimator $\widehat{\theta}$, it is not necessary for every boundary point (x, y, z) to satisfy $x \geq a$, $y \geq b$ and $z \geq c$. As $n \geq a + b + c$, the point (a, b, c) is an accessible/boundary point and the estimator is positive for some boundary points. For example, if we consider the trinomial sampling plan $x + y + z = 15$, and consider estimation of $\theta = p^2q^3r^4$, then, at the boundary point $\alpha_{(0,0,0)} = (2, 10, 3)$, the estimate is $\hat{\theta} = 0$. In this case, the question that naturally arises is that whether we can provide an unbiased estimator for $\theta = p^2q^3r^4$ based on the mentioned trinomial plan. We note that

$$\begin{aligned}\theta &= p^2q^3r^4 = p^2q^3r^4(p+q+r)^6\\ &= \sum_{u=0}^{6} \sum_{\substack{v=0\\ u+v+w=6}}^{6} \sum_{w=0}^{6} \frac{6!}{u!v!w!} p^{2+u}q^{3+v}r^{4+w}.\end{aligned}$$

Each of the 7 terms in the expansion above hits a boundary point (i.e., $x + y + z = 15$) and hence provides an estimate of the corresponding probability term by the use of an indicator function. For example, we find $p^4q^5r^6$ with coefficient $(6!/2!2!2!) = 90$, whereas the binomial expansion of $(p + q + r)^{15}$ yields its coefficient as $(15!/4!5!6!) = 3 \times 7 \times 11 \times 13 \times 14 \times 15 = 630{,}630 >> 90$. Hence all the terms can be estimated.

Next, let us consider the inverse trinomial sampling plan $P_{(t,0,0)}$ with $x = t$; $y, z = 0, 1, 2, \ldots, t$ being a generic constant. The boundary point is denoted as $\alpha_{(0,0,0)} = (t, y, z)$ for such sampling plan. We use pull-down sampling plan 2-step along X-axis, 3-step along Y-axis, and 4-step along Z-axis. Then the new boundary point is $\alpha_{(-2,-3,-4)} = (t-2, y-3, z-4)$. Using path-counting formula we obtain $k(\alpha_{(0,0,0)}) = \frac{(t+y+z-1)!}{(t-1)!y!z!}$ and $k(\alpha_{(-2,-3,-4)}) = \frac{(t+y+z-10)!}{(t-3)!(y-3)!(z-4)!}$. Therefore, estimator of $\theta = p^2q^3r^4$ is

$$\hat{\theta} = \frac{k(\alpha_{(-2,-3,-4)})}{k(\alpha_{(0,0,0)})} = \frac{\frac{(t+y+z-10)!}{(t-3)!(y-3)!(z-4)!}}{\frac{(t+y+z-1)!}{(t-1)!y!z!}} = \frac{(t-1)_{(2)}(y)_{(3)}(z)_{(4)}}{(t+y+z-1)_{(9)}}.$$

Suppose we have $x = t = 5$, and $y = 5$, $z = 5$. Here we pull down the sampling plan 2-step along X-axis, 3-step along Y-axis, and 4-step along Z-axis. Then, the new boundary point is $\alpha_{(-2,-3,-4)} = (3, 2, 1)$. Therefore, the estimate of θ at $\alpha_{(0,0,0)} = (5, 5, 5)$ is

$$\hat{\theta} = \frac{k(\alpha_{(-2,-3,-4)})}{k(\alpha_{(0,0,0)})} = \frac{\frac{5!}{2!2!1!}}{\frac{14!}{4!5!5!}} = 0.000119.$$

Remark 2.5.3 With respect to the inverse trinomial sampling plan $x = t$, for the existence of the estimator $\hat{\theta}$, it is not necessary to have $y \geq b$ and $z \geq c$. However, the plan parameter 't' should be such that $t + y + z \geq a + b + c$.

2.5.4 *Unbiased Estimation of $\frac{1}{pq}$ and $\frac{1}{pqr}$ under an Experiment Performed until (a, a, a) is achieved, a being a Specified Positive Integer*

Here we perform an experiment with three possible outcomes with probabilities p, q, and r, until we get $x = a$, $y = a$, $z = a$, where x, y, and z are as defined before. For simplicity of understanding, let us take $a = 4$. The boundary points of the sampling plan (4, 4, 4) lie on the red lines in Fig. 2.15.

It is readily seen that the boundary points of the plan are (i) (4, 4, 4); (ii) (x, 4, 4), $x \geq 5$; (iii) (4, y, 4), $y \geq 5$; and (iv) (ii) (4, 4, z), $z \geq 5$.

We leave it to the readers to show that the plan is closed.

To reach the point (4, 4, 4) from (0, 0, 0), we move via (3, 4, 4) or (4, 3, 4) or (4, 4, 3). Hence, by the path-counting formula, the value of k((4, 4, 4)) is $3 \times \frac{11!}{3!4!4!} =$ 34,650.

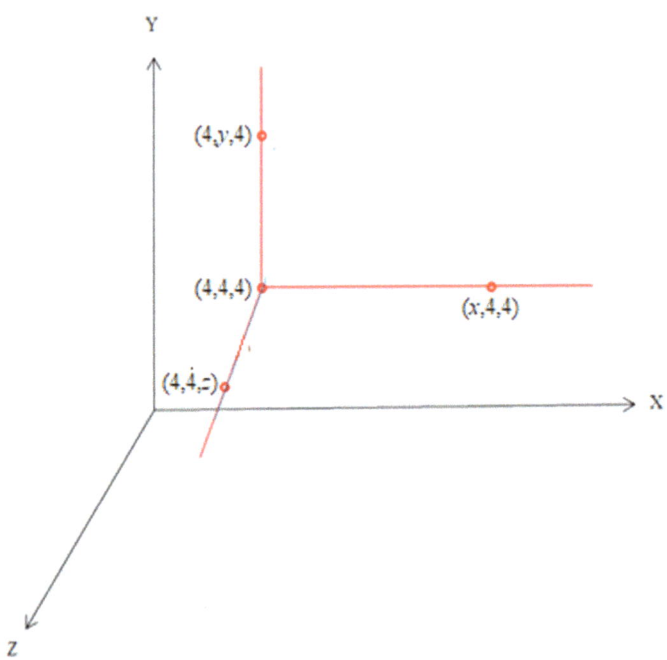

Fig. 2.15 Sampling plan (4, 4, 4)

For the type (ii) boundary points, let us denote a typical boundary point as $\alpha_{(0,0,0)} = (x, 4, 4)$. This point can be reached only through the point $(x, 3, 4)$ or $(x, 4, 2)$. So, in this case, the value of $k\left(\alpha_{(0,0,0)}\right)$ is $2 \times \frac{(x+7)!}{x!3!4!}$.

Again, for the type (iii) boundary points, a typical boundary point is $\alpha_{(0,0,0)} = (4, y, 4)$, and it can be reached only through the point $(3, y, 4)$ or $(4, y, 3)$. Hence, the value of $k\left(\alpha_{(0,0,0)}\right)$ is $2 \times \frac{(y+7)!}{3!y!4!}$.

Finally, for type (iv) boundary points, a typical boundary point has the form $\alpha_{(0,0,0)} = (4, 4, z)$, which can be reached only through the point $(4, 3, z)$ or $(3, 4, z)$. So, the value of $k\left(\alpha_{(0,0,0)}\right)$ is $2 \times \frac{(z+7)!}{3!4!z!}$.

To estimate $\frac{1}{pq}$, we use 1-step push-up plan along the X-axis and 1-step push-up plan along the Y-axis, simultaneously. This results in the plan $P_{(+1,+1,0)}$ with boundary points $\alpha_{(+1,+1,0)}$, which are of the form (i′) (5, 5, 4), (ii′) $(x, 5, 4)$; $x \geq 6$, (iii′) $(5, y, 4)$; $y \geq 6$, and (iv′) $(5, 5, z)$; $z \geq 5$. On the other hand, to estimate $\frac{1}{pqr}$, we push-up one step along X-axis, Y-axis, and Z-axis, simultaneously. This gives the new boundary points $\alpha_{(+1,+1,+1)}$, which are (i″) (5, 5, 5), (ii″) $(x, 5, 5)$; $x \geq 6$, (iii″) $(5, y, 5)$; $y \geq 6$, and (iv″) $(5, 5, z)$; $z \geq 6$. The boundary points and the unbiased estimates have been indicated in Table 2.8.

Remark 2.5.4 The above discussions on trinomial distribution have been compiled from Tarafdar et al. (2020). For detailed understanding of parameter estimation related to a trinomial distribution, the readers are referred to the paper.

2.6 Unbiased Estimation of Parameters in a Tetranomial Set-Up

Consider an experiment which has four possible outcomes, namely O_1, O_2, O_3, and O_4, with probabilities $p^2, q^2, r^2, 2(pq + pr + qr)$, respectively, such that $0 < p, q, r < p + q + r = 1$. We shall discuss the problems of (i) unbiased estimation of p; and (ii) unbiased estimation of p, q, and r simultaneously.

For simultaneous estimation of p, q, and r, without loss of generality we may concentrate on simultaneous estimation of p and q only. This is because the estimate of r will follow from the restriction $p + q + r = 1$. As the probabilities of the outcomes O_3 and O_4 only depend on r, it seems logical to combine the last two outcomes together as a single outcome O_0, say, and focus on three outcomes O_1, O_2, and O_0 with probabilities p^2, q^2 and $1 - \left(p^2 + q^2\right) = r^2 + 2(pq + pr + qr)$, respectively.

It is very clear that no finite sample size strategy will suffice for unbiased estimation of the parameters in the exact sense. We, therefore, proceed to suggest a sequential sampling procedure for estimating the parameters.

Table 2.8 Estimation of $\frac{1}{pq}$ and $\frac{1}{pqr}$

$\alpha_{(0,0,0)}$	$\alpha_{(+1,+1,0)}$	$\alpha_{(+1,+1,+1)}$	$k(\alpha_{(0,0,0)})$	$k(\alpha_{(+1,+1,0)})$	$k(\alpha_{(+1,+1,+1)})$	$\widehat{\frac{1}{pq}} = \frac{k(\alpha_{(+1,+1,0)})}{k(\alpha_{(0,0,0)})}$	$\widehat{\frac{1}{pqr}} = \frac{k(\alpha_{(+1,+1,+1)})}{k(\alpha_{(0,0,0)})}$
$(4, 4, 4)$	$(5, 5, 4)$	$(5, 5, 5)$	$3 \times \frac{11!}{3!4!4!}$	$\frac{13!}{5!5!3!} + 2 \times \frac{13!}{5!4!4!}$	$3 \times \frac{14!}{5!5!4!}$	7.28	21.84
$(x, 4, 4); x \geq 5$	$(x, 5, 4); x \geq 6$	$(x, 5, 5); x \geq 6$	$2 \times \frac{(x+7)!}{x!3!4!}$	$\frac{(x+8)!}{x!5!3!} + \frac{(x+8)!}{x!4!4!}$	$2 \times \frac{(x+9)!}{x!4!5!}$	$\frac{9(x+8)}{40}$	$\frac{(x+8)(x+9)}{20}$
$(4, y, 4); y \geq 5$	$(5, y, 4); y \geq 6$	$(5, y, 5); y \geq 6$	$2 \times \frac{(y+7)!}{3!y!4!}$	$\frac{(y+8)!}{5!y!3!} + \frac{(y+8)!}{4!y!4!}$	$2 \times \frac{(y+9)!}{4!y!5!}$	$\frac{9(y+8)}{40}$	$\frac{(y+8)(y+9)}{20}$
$(4, 4, z); z \geq 5$	$(5, 5, z); z \geq 5$	$(5, 5, z); z \geq 6$	$2 \times \frac{(z+7)!}{3!4!z!}$	$2 \times \frac{(z+9)!}{4!5!z!}$	$2 \times \frac{(z+9)!}{4!5!z!}$	$\frac{(z+8)(z+9)}{20}$	$\frac{(z+8)(z+9)}{20}$

2.6.1 Sequential Sampling Plan

Let us define $\theta_1 = p^2, \theta_2 = q^2, \theta_0 = r^2 + 2(pq + pr + qr)$. Then,

$$p = \sqrt{\theta_1}$$

$$q = \sqrt{\theta_2}$$

and $\theta_1 + \theta_2 + \theta_0 = 1$,

Since $p + q + r = 1$.

As we wish to estimate all the three parameters p, q, and r, we need the representation of all the three outcomes O_1, O_2, and O_0 in the sampling scheme. We, therefore, define our sampling plan as follows:

Continue sampling from the trinomial distribution until each of the three outcomes O_1, O_2, and O_0 is represented 'at least once'.

Clearly, the plan indicates termination of sampling when the frequencies of the 3 outcomes are as given in Table 2.9.

Here, $r_1, r_2, t_2, t_3, k_1, k_3 \geq 1$.

As indicated in Table 2.9, there can be three types of frequencies. The sampling procedure terminates with

(i) Frequency Type I, if the outcome O_1 occurs for the first time at the end, while the outcomes O_2 and O_0 are observed at least once in the trials before that;
(ii) Frequency Type II, if the outcome O_2 is observed for the first time at the end, and the other two outcomes O_1 and O_0 have occurred at least once in the trials before that;
(iii) Frequency Type III, if the outcome C_0 occurs for the first time at the end, while O_1 and O_2 occur at least once before that.

It is readily seen that the three frequency types are mutually exclusive, and it is also easy to verify that

$$\begin{aligned} P_1 &= P\big[\text{Termination of sampling via Frequency Type I}\big] \\ &= \frac{(1+\theta_1)\theta_2\theta_0}{(1-\theta_2)(1-\theta_0)} \end{aligned}$$

$$P_2 = P\big[\text{Termination of sampling via Freq. Type II}\big]$$

Table 2.9 Frequencies of the three outcomes O_1, O_2, and O_0 in the sample

Outcome	Freq. (Type I)	Freq. (Type II)	Freq. (Type III)
C_0	r_1	r_2	1
C_1	1	t_2	t_3
C_2	k_1	1	k_3

$$= \frac{(1+\theta_2)\theta_1\theta_0}{(1-\theta_1)(1-\theta_0)}$$

$$P_3 = P\left[\text{Termination of sampling via Freq. Type III}\right]$$
$$= \frac{(1+\theta_2)\theta_1\theta_0}{(1-\theta_1)(1-\theta_0)}.$$

Further, $P_1 + P_2 + P_3 = 1$ which indicates that the three frequency types are collectively exhaustive.

2.6.2 Unbiased Estimation of Parameters

We first consider the problem of estimating p.

We can write

$$p = \sqrt{\theta_1} = \theta_1 \times \theta_1^{-1/2} = \theta_1[1 - (\theta_2 + \theta_0)]^{-1/2}$$
$$= \theta_1 \sum_{j=0}^{\infty} a_j(\theta_2 + \theta_0)^j,$$

where $a_j = \frac{(2j)!}{2^{2j}(j!)^2}$, $j = 0, 1, 2, \ldots$.

Clearly, $a_0 = 1$ and $a_j < 1$ for $j = 1, 2, 3, \ldots$. Further, a_j's form a decreasing sequence, and $a_j + a_k < 1$ for any $j, k \geq 1, j \neq k$.

For varying j, we get the following events and their probabilities:

(i) O_1 appears in the very first trial. This has probability θ_1, and it corresponds to '$j = 0$'.
(ii) O_1 appears for the first time in the second trial; that is, O_2 or O_0 appears in the first trial. The probability for this event is $\theta_1(\theta_2 + \theta_0)$. This corresponds to '$j = 1$'.
(iii) O_1 appears for the first time in the third trial; that is, O_2 and/or O_0 appear(s) in the first two trials. The probability for this event is $\theta_1(\theta_2 + \theta_0)^2$, and it corresponds to '$j = 2$'; and so on.

Remark 2.6.1 It may be noted that for the unbiased estimation of 'p' alone, we have used only two outcomes: O_1 and its compliment. Likewise, we may develop a sequential sampling plan for unbiased estimation of 'q' alone. Thus, in principle, this calls for two separate plans. A non-trivial proposal would be to suggest only one plan which can provide simultaneous unbiased estimation of both 'p' and 'q', and, hence, of 'r'. Toward this we have proposed the above three-outcome sequential sampling scheme. We shall analyze the above scheme in details.

2.6.2.1 Classification of Samples

To estimate p, we collect outcomes of independent trials till O_1 appears. However, this may result in not observing either O_2 or O_0, or both, before observing O_1. This means we need to continue sampling as per our sampling procedure described in Sect. 6.1, though we shall only use the data till O_1 appears. As a result, we may have complete or incomplete data set, which we may classify as a terminating sample or a non-terminating sample. The definitions of such samples are as below.

Terminating sample: A terminating sample is one in which each of the outcomes O_2 and O_0 appears at least once before the sampling ends with the appearance of the outcome O_1. Examples of terminating samples are: $\{(O_2, O_0, O_1), (O_0, O_2, O_1)\}$ in 3 trials, $\{(O_2, O_2, O_0, O_1), (O_2, O_0, O_2, O_1), (O_0, O_2, O_2, O_1), (O_0, O_0, O_2, O_1), (O_0, O_2, O_0, O_1), (O_2, O_0, O_0, O_1)\}$ in 4 trials, etc. In general, $\{(O_2(k), O_0(r), O_1) \mid k + r = j \geq 2\}$ are terminating samples, where $[O_2(k)\ O_0(r)]$ refers to all possible $(2^{k+r} - 2)$ combinations of O_2 and O_0 in $(k + r)$ trials in which O_2 appears k times and O_0 appears r times, prior to appearance of O_1.

Non-terminating sample: A sample in which none or only one of the two outcomes O_2 and O_0 appears before the appearance of O_1 is a non-terminating sample. For example, $\{(O_1; ...)\}$, $\{(O_2, O_1; ...), (O_0, O_1; ...)\}$, $\{(O_2, O_2, O_1; ...), (O_0, O_0, O_1; ...)\}$, etc. are all non-terminating samples. In general, $\{(O_2(k), O_1); ... \mid k \geq 1\}$ and $\{(O_0(r), O_1); ... \mid r \geq 1\}$ are non-terminating samples, where $O_i(j)$ indicates the appearance of O_i in the first j trials before O_1 appears. We use '...' to indicate that the sample is incomplete.

Remark 2.6.2 It is noteworthy that the non-terminating samples of the type $(O_2(k), O_1;...)$ or $(O_0(r), O_1; ...)$ do terminate with probability one, conditional on the first part. The idea is to identify the nature of the 'complete' sample by referring to the first part (observed) and then assigning an indicator to it.

For any sample S, let $I(S)$ denote an indicator variable such that

$$\begin{aligned} I(S) &= 1, \quad \text{if the first trial yields } O_1 \\ &= 0, \quad \text{otherwise,} \end{aligned}$$

and, for $j = 1, 2, \ldots$,

$$\begin{aligned} I(S) &= a_j, \quad \text{if in the first } j \text{ trials } O_2 \text{ and/or } O_0 \text{ appear(s) followed by } O_1 \text{ at the } (j+1)\text{st trial} \\ &= 0, \quad \text{otherwise.} \end{aligned}$$

Table 2.10 gives all information toward developing $\hat{p}$, based on terminating samples.

Likewise, we indicate the information obtained from the non-terminating samples in Table 2.11.

Table 2.10 Information based on terminating samples

j	Sample (S)	$I(S)$
2	$\{(O_2, O_0, O_1), (O_0, O_2, O_1)\}$	3/8 for each sample
3	$\{(C_2, C_2, C_0, C_1), (C_2, C_0, C_2, C_1), (C_0, C_2, C_2, C_1), (C_0, C_0, C_2, C_1), (C_0, C_2, C_0, C_1), (C_2, C_0, C_0, C_1)\}$	5/16 for each of the 6 samples
$j\,(j \geq 4)$	$\{(C_2(k), C_0(r), C_1) \mid k + r = j \geq 4\}$	$\frac{(2j)!}{2^{2j}(j!)^2}$ for each of the $2(2^{j-1} - 1)$ samples

Table 2.11 Information based on non-terminating samples

j	Sample (S)	$I(S)$
0	$(O_1; \ldots)$	1
1	$\{(O_2, O_1; \ldots), (O_0, O_1; \ldots)\}$	1/2 for each sample
2	$\{(O_2, O_2, O_1; \ldots), (O_0, O_0, O_1; \ldots)\}$	3/8 for each sample
3	$\{(O_2, O_2, O_2, O_1; \ldots), (O_0, O_0, O_0, O_1; \ldots)\}$	5/16 for each sample
$j\,(j \geq 4)$	$\{(O_2(j), O_1; \ldots), (O_0(j), O_1; \ldots) \mid j \geq 4\}$	$\frac{(2j)!}{2^{2j}(j!)^2}$ for each sample

2.6.2.2 Unbiased Estimation of *p*

For a given sample S, we define an estimator of p as

$$\begin{aligned} \hat{p} &= 1, \quad \text{with probability } I(S) \\ &= 0, \quad \text{with probability } 1 - I(S). \end{aligned}$$

To obtain the expectation of $\hat{p}$, consider the following notations:

T $\equiv$ set of terminating samples
NT $\equiv$ set of non-terminating samples
$s_{ij}^{\mathrm{T}} =$ ith terminating sample corresponding to j, $i = 1, 2, \ldots, 2^j - 2$, $j = 2, 3, \ldots$.
$s_{ij}^{\mathrm{NT}} =$ ith non-terminating sample corresponding to j, $i = 1, 2, \ldots, b_j$, $j = 0, 1, 2, \ldots$.

$$\begin{aligned} b_j &= 1 \quad \text{if } j = 0 \\ &= 2 \quad \text{if } j \geq 1. \end{aligned}$$

Then,

$$E(\hat{p}) = \sum_{s \in \mathrm{T}} I(s) \Pr[S = s] + \sum_{s \in \mathrm{NT}} I(s) \Pr[S = s]$$

$$= \sum_{j=2}^{\infty}\sum_{i=1}^{2^j-2} I\left(s_{ij}^{\mathrm{T}}\right)\Pr\left[S = s_{ij}^{\mathrm{T}}\right] + \sum_{j=0}^{\infty}\sum_{i=1}^{b_j} I\left(s_{ij}^{\mathrm{NT}}\right)\Pr\left[S = s_{ij}^{\mathrm{T}}\right]. \tag{2.6.1}$$

Let us look at terminating and non-terminating samples separately.
For terminating samples, we have, for $j \geq 2$,

$$\sum_{i=1}^{2^j-2} I\left(s_{ij}^{\mathrm{T}}\right)\Pr\left[S = s_{ij}\right] = a_j \Pr\left[(O_2(k), O_0(r), O_1) | k + r = j; k, r \geq 1\right] \tag{2.6.2}$$

For non-terminating samples, we have

(i) for $j = 0$,

$$\sum_{i=1}^{b_0} I\left(s_{i0}^{\mathrm{NT}}\right)\Pr\left[S = s_{i0}^{\mathrm{NT}}\right] = I(s_{10})\Pr[(O_1)] = 1 \times \theta_1 = \theta_1 \tag{2.6.3}$$

(ii) for $j = 1$,

$$\sum_{i=1}^{b_1} I\left(s_{i1}^{\mathrm{NT}}\right)\Pr\left[S = s_{i1}^{\mathrm{NT}}\right] = I(s_{10})\Pr[(O_2, O_1)]$$
$$+ I(s_{20})\Pr[(O_0, O_1)] = (1/2)(\theta_2 + \theta_0)\theta_1 \tag{2.6.4}$$

(iii) for $j \geq 2$,

$$\sum_{i=1}^{b_j} I\left(s_{ij}^{\mathrm{NT}}\right)\Pr\left[S = s_{ij}^{\mathrm{NT}}\right] = a_j \Pr\left[(O_2(j), O_1) + a_j \Pr\left[(O_2(j), O_1)\right]. \tag{2.6.5}$$

So, for $j \geq 2$,

$$\begin{aligned}(2.6.2) + (2.6.5) &= a_j \Pr\left[\text{the first } j \text{ trials give } O_2 \text{ and/or } O_0, \text{ followed by } O_1\right] \\ &= a_j(\theta_2 + \theta_0)^j \theta_1.\end{aligned}$$

Hence, (2.6.1) gives

$$E(\hat{p}) = \sum_{j=0}^{\infty} a_j(\theta_2 + \theta_0)^j \theta_1 = p, \text{ whatever be the true values of } (p, q).$$

Thus, $\hat{p}$ is an unbiased estimator of p.

Remark 2.6.3 For unbiased estimation of 'q', we reverse the roles of outcomes 'O_1' and 'O_2' in the trinomial set-up with outcomes O_1, O_2, and O_0. The procedure remains the same as before. The non-terminating samples of the type $\{(O_2; \ldots); (O_1, O_2; \ldots), (O_0, O_2; \ldots)$, etc., are suitably used to get $\hat{q}$.

We next reproduce a simulation study done by Sultana et al. (2023) to show how the method works. In fact, all the simulation studies indicated henceforth are reproduced from Sultana et al. (2023).

Suppose $p = 0.3$, $q = 0.5$. An experiment with repeated trials is performed till O_1 appears. Nine trials had to be conducted for the purpose, and the outcomes came out as follows:

Trial No.	1	2	3	4	5	6	7	8	9
Outcome	O_2	O_2	O_0	O_0	O_2	O_0	O_0	O_0	O_1

So, here $j = 8$, and we have a terminating sample. Let S denote the sample.

Now, $a_8 = 0.1964$ and therefore, $I(S) = 0.1964$.

We next perform a binary trial [designated as *auxiliary known randomization*] with success probability 0.1964. The outcome is a 'failure'. Therefore, $\hat{p} = 0$.

In the same sample S, we observe that O_2 appears in the very first trial. So, for estimation of q, we have $j = 0$, which gives a non-terminating sample. Then, $I(S) = 1$, and we obtain $\hat{q} = 1$.

We may avoid such indicator-based estimates by independently repeating the above exercise some 'N' times and taking the average score.

Consider $p = 0.30$, $q = 0.5$, and $N = 20$. Table 2.12 summarizes the results obtained through simulation.

Hence, the estimate of p is $\hat{\hat{p}} = \frac{6}{20} = 0.30$.

The above procedure was repeated 100 times, and the estimated standard error of the estimate came out to be 0.09709.

2.6.2.3 Simultaneous Estimation of p and q

For simultaneous estimation of p and q, we note that the unbiased estimators are based on $I(S)$, which are basically a_j-values, depending on the desired outcomes.

For $\hat{p}$, we note the appearance of outcome O_1 for the first time while for $\hat{q}$, we go for the first time appearance of O_2. A simulation study, with $p = 0.3$, $q = 0.5$, $r = 0.2$, is carried out with 20 repetitions of the experiment. The outcomes are given in Table 2.13. In the table, j denotes the number of trials before the first appearance of O_1, while k denotes the number of trials before the first appearance of O_2.

Here, the estimates are $\hat{\hat{p}} = 0.30$, $\hat{\hat{q}} = 0.55$. Hence, the estimate of r is $\hat{\hat{r}} = 0.15$.

The above procedure was repeated 100 times, and the estimated standard errors were obtained as $SE\left(\hat{\hat{p}}\right) = 0.01357735$ and $SE\left(\hat{\hat{q}}\right) = 0.01593257$.

We recall from Sect. 2.6.2 that a_j's form a decreasing sequence, and $a_j + a_k < 1$ for any $j, k \geq 1, j \neq k$, where $0 < a_j < 1$ for $j = 1, 2, \ldots$.

Table 2.12 Outcomes of 20 repetitions of the experiment

Serial No.	Type of sample	j	a_j	$\hat{p}$
1	T	12	0.16118026	0
2	T	15	0.14446445	0
3	NT	0	1.00000000	1
4	T	3	0.31250000	0
5	T	8	0.19638062	0
6	T	11	0.16818810	0
7	T	8	0.19638062	0
8	T	10	0.17619705	0
9	T	9	0.18547058	0
10	T	15	0.14446445	0
11	T	5	0.24609375	0
12	T	3	0.31250000	1
13	T	2	0.37500000	1
14	T	12	0.16118026	0
15	T	15	0.14446445	0
16	T	14	0.14944598	1
17	NT	1	0.50000000	1
18	T	16	0.13994993	0
19	T	2	0.37500000	0
20	NT	0	1.00000000	1

T: terminating, *NT*: non-terminating

So, given a sample, terminating or non-terminating, we can identify 'j' for O_1 and 'k' for O_2. Further, as $a_j + a_k < 1$, we can perform a 3-outcometrial with probabilities $(a_j, a_k, 1 - a_j - a_k)$ and end up with an output like (1, 0, 0), (0, 1, 0), or (0, 0, 1), with an obvious interpretation. The different possible samples and the corresponding outputs, along with the probabilities, are as given below.

Sample	Output	Probability
$(O_1; \ldots)$	$(1, 0, 0)$	1
$(O_2; \ldots)$	$(0, 1, 0)$	1
O_0 followed by O_1/O_2 for termination	$(1, 0, 0)$	a_j, if j combinations of O_0 and O_2 before first appearance of O_1
	$(0, 1, 0)$	a_k, if k combinations of O_0 and O_1 before first appearance of O_2
	$(0, 0, 1)$	$1 - a_j - a_k$, if j O_1's and k O_2's before terminating with O_1 or O_2

Table 2.13 Outcomes of 20 repetitions of the experiment

Serial No.	j	k	a_j	a_k	$\hat{p}$	$\hat{q}$
1	5	3	0.24609375	0.3125000	1	0
2	1	0	0.50000000	1.0000000	1	1
3	3	2	0.31250000	0.3750000	0	1
4	4	0	0.27343750	1.0000000	0	1
5	2	4	0.37500000	0.2734375	1	0
6	7	15	0.20947266	0.1444644	0	0
7	1	2	0.50000000	0.3750000	1	0
8	24	0	0.11456650	1.0000000	0	1
9	7	10	0.20947266	0.1761971	0	0
10	4	5	0.27343750	0.2460938	1	0
11	13	0	0.15498102	1.0000000	0	1
12	18	2	0.13206060	0.3750000	0	0
13	37	2	0.09243938	0.3750000	0	1
14	2	1	0.37500000	0.5000000	1	1
15	11	2	0.16818810	0.3750000	0	0
16	3	1	0.31250000	0.5000000	0	0
17	3	0	0.31250000	1.0000000	0	1
18	7	3	0.20947266	0.3125000	0	1
19	7	1	0.20947266	0.5000000	0	1
20	3	0	0.31250000	1.0000000	0	1

Let 'N' repetitions of the experiment be conducted. Then, averaging over the outputs would yield proper estimates.

A simulation study is carried out taking $p = 0.3$, $q = 0.5$, $r = 0.2$. Hundred repetitions of the trinomial experiment were carried out, and the outcomes obtained are presented in Table 2.14.

So the estimates are $\hat{\hat{p}} = 0.28$, $\hat{\hat{q}} = 0.45$ and $\hat{\hat{r}} = 0.27$.

Table 2.15 gives the estimates of p and q obtained from 10 sets of 100 repetitions of the experiment for different sets of values of (p, q).

Remark 2.6.4 For simultaneous unbiased estimation of p and q, sequential samples have been generated with appropriately defined stopping rules. The estimates obtained are found to be considerably close to the true parameter values.

Remark 2.6.5 All the simulation studies indicated in Sect. 2.6 are reproduced from Sultana et al. (2023). The mentioned study is fairly recent, and it encompasses many aspects of simulation. However, most readers may not have easy access to the reference journal [sequential analysis]. So, we decided to re-use the same simulation study to provide necessary explanations.

Table 2.14 Outcomes from 100 repetitions of the trinomial experiment

Serial No.	j	k	a_j	a_k	$1 - a_j - a_k$	$\hat{p}$	$\hat{q}$	$\hat{r}$
1	7	1	0.2094727	0.5000000	0.2905273	0	0	1
2	5	6	0.2460938	0.2255859	0.5283203	1	0	0
3	17	11	0.1358338	0.1681881	0.6959781	0	0	1
4	14	2	0.1494460	0.3750000	0.4755540	0	0	1
5	37	2	0.0924394	0.3750000	0.5325606	1	0	0
6	15	2	0.1444644	0.3750000	0.4805356	0	1	0
7	3	1	0.3125000	0.5000000	0.1875000	0	1	0
8	–	0	0.0000000	1.0000000	0.0000000	0	1	0
9	2	6	0.3750000	0.2255859	0.3994141	1	0	0
10	3	6	0.3125000	0.2255859	0.4619141	1	0	0
11	9	2	0.1854706	0.3750000	0.4395294	0	0	1
12	3	2	0.3125000	0.3750000	0.3125000	0	0	1
13	5	8	0.2460938	0.1963806	0.5575256	0	0	1
14	0	–	1.0000000	0.0000000	0.0000000	1	0	0
15	2	3	0.3750000	0.3125000	0.3125000	1	0	0
16	12	2	0.1611803	0.3750000	0.4638197	1	0	0
17	22	2	0.1196042	0.3750000	0.5053958	0	1	0
18	–	0	0.0000000	1.0000000	0.0000000	0	1	0
19	0	–	1.0000000	0.0000000	0.0000000	1	0	0
20	3	2	0.3125000	0.3750000	0.3125000	0	0	1
21	11	2	0.1681881	0.3750000	0.4568119	0	0	1
22	29	7	0.1043168	0.2094727	0.6862106	1	0	0
23	0	–	1.0000000	0.0000000	0.0000000	1	0	0
24	1	6	0.5000000	0.2255859	0.2744141	0	1	0
25	11	4	0.1681881	0.2734375	0.5583744	0	0	1
26	0	–	1.0000000	0.0000000	0.0000000	1	0	0
27	1	2	0.5000000	0.3750000	0.1250000	0	1	0
28	4	3	0.2734375	0.3125000	0.4140625	0	0	1
29	–	0	0.0000000	1.0000000	0.0000000	0	1	0
30	31	1	0.1009237	0.5000000	0.3990763	0	1	0
31	14	5	0.1494460	0.2460938	0.6044603	0	1	0
32	0	–	1.0000000	0.0000000	0.0000000	1	0	0
33	14	1	0.1494460	0.5000000	0.3505540	0	1	0
34	–	0	0.0000000	1.0000000	0.0000000	0	1	0

(continued)

Table 2.14 (continued)

Serial No.	j	k	a_j	a_k	$1 - a_j - a_k$	$\hat{p}$	$\hat{q}$	$\hat{r}$
35	2	4	0.3750000	0.2734375	0.3515625	1	0	0
36	21	1	0.1223857	0.5000000	0.3776143	0	1	0
37	–	0	0.0000000	1.0000000	0.0000000	0	1	0
38	16	5	0.1399499	0.2460938	0.6139563	1	0	0
39	10	1	0.1761971	0.5000000	0.3238029	0	1	0
40	0	–	1.0000000	0.0000000	0.0000000	1	0	0
41	–	0	0.0000000	1.0000000	0.0000000	0	1	0
42	18	1	0.1320606	0.5000000	0.3679394	0	0	1
43	43	43	0.0000000	1.0000000	0.0000000	0	1	0
44	–	0	0.0000000	1.0000000	0.0000000	0	1	0
45	10	2	0.1761971	0.3750000	0.4488029	0	0	1
46	5	3	0.2460938	0.3125000	0.4414063	1	0	0
47	–	0	0.0000000	1.0000000	0.0000000	0	1	0
48	15	2	0.1444644	0.3750000	0.4805356	0	0	1
49	27	7	0.1080768	0.2094727	0.6824505	0	0	1
50	0	–	1.0000000	0.0000000	0.0000000	1	0	0
51	16	2	0.1399499	0.3750000	0.4850501	0	0	1
52	–	0	0.0000000	1.0000000	0.0000000	0	1	0
53	1	2	0.5000000	0.3750000	0.1250000	1	0	0
54	3	7	0.3125000	0.2094727	0.4780273	0	0	1
55	–	0	0.0000000	1.0000000	0.0000000	0	1	0
56	1	5	0.5000000	0.2460938	0.2539063	0	1	0
57	4	6	0.2734375	0.2255859	0.5009766	0	1	0
58	10	2	0.1761971	0.3750000	0.4488029	0	1	0
59	23	5	0.1170041	0.2460938	0.6369022	0	1	0
60	–	0	0.0000000	1.0000000	0.0000000	0	1	0
61	2	6	0.3750000	0.2255859	0.3994141	0	1	0
62	–	0	0.0000000	1.0000000	0.0000000	0	1	0
63	2	6	0.3750000	0.2255859	0.3994141	1	0	0
64	8	2	0.1963806	0.3750000	0.4286194	0	1	0
65	11	4	0.1681881	0.2734375	0.5583744	0	0	1
66	2	1	0.3750000	0.5000000	0.1250000	1	0	0
67	3	5	0.3125000	0.2460938	0.4414063	0	0	1
68	–	0	0.0000000	1.0000000	0.0000000	0	1	0

(continued)

Table 2.14 (continued)

Serial No.	j	k	a_j	a_k	$1 - a_j - a_k$	$\hat{p}$	$\hat{q}$	$\hat{r}$
69	5	7	0.2460938	0.2094727	0.5444336	0	0	1
70	–	0	0.0000000	1.0000000	0.0000000	0	1	0
71	3	10	0.3125000	0.1761971	0.5113029	0	1	0
72	29	2	0.1043168	0.3750000	0.5206832	0	1	0
73	32	7	0.0993468	0.2094727	0.6911806	0	1	0
74	0	–	1.0000000	0.0000000	0.0000000	1	0	0
75	9	3	0.1854706	0.3125000	0.5020294	1	0	0
76	16	2	0.1399499	0.3750000	0.4850501	0	0	1
77	1	2	0.5000000	0.3750000	0.1250000	1	0	0
78	15	1	0.1444644	0.5000000	0.3555356	0	1	0
79	0	–	1.0000000	0.0000000	0.0000000	1	0	0
80	4	1	0.2734375	0.5000000	0.2265625	0	0	1
81	10	2	0.1761971	0.3750000	0.4488029	0	1	0
82	–	0	0.0000000	1.0000000	0.0000000	0	1	0
83	4	14	0.2734375	0.1494460	0.5771165	0	1	0
84	14	2	0.1494460	0.3750000	0.4755540	0	0	1
85	10	13	0.1761971	0.1549810	0.6688219	0	0	1
86	5	1	0.2460938	0.5000000	0.2539063	0	1	0
87	2	3	0.3750000	0.3125000	0.3125000	1	0	0
88	10	1	0.1761971	0.5000000	0.3238029	0	0	1
89	16	7	0.1399499	0.2094727	0.6505774	0	1	0
90	–	0	0.0000000	1.0000000	0.0000000	0	1	0
91	0	–	1.0000000	0.0000000	0.0000000	1	0	0
92	7	8	0.2094727	0.1963806	0.5941467	0	0	1
93	7	1	0.2094727	0.5000000	0.2905273	0	1	0
94	5	9	0.2460938	0.1854706	0.5684357	1	0	0
95	–	0	0.0000000	1.0000000	0.0000000	0	1	0
96	8	21	0.1963806	0.1223857	0.6812337	1	0	0
97	6	1	0.2255859	0.5000000	0.2744141	0	1	0
98	4	1	0.2734375	0.5000000	0.2265625	0	0	1
99	2	3	0.3750000	0.3125000	0.3125000	0	0	1
100	–	0	0.0000000	1.0000000	0.0000000	0	1	0

'–' indicates not observed

Table 2.15 Estimates of p and q from 10 sets of 100 repetitions of the experiment using 36 different sets of values of (p, q)

No.	p	q	$\hat{\hat{p}}$	$\hat{\hat{q}}$
1	0.1	0.1	0.110	0.104
2	0.1	0.2	0.113	0.179
3	0.1	0.3	0.113	0.279
4	0.1	0.4	0.093	0.398
5	0.1	0.5	0.098	0.485
6	0.1	0.6	0.066	0.610
7	0.1	0.7	0.049	0.711
8	0.1	0.8	0.046	0.802
9	0.2	0.1	0.186	0.116
10	0.2	0.2	0.184	0.222
11	0.2	0.3	0.189	0.276
12	0.2	0.4	0.183	0.372
13	0.2	0.5	0.161	0.444
14	0.2	0.6	0.145	0.571
15	0.2	0.7	0.106	0.685
16	0.3	0.1	0.317	0.097
17	0.3	0.2	0.271	0.184
18	0.3	0.3	0.274	0.309
19	0.3	0.4	0.256	0.369
20	0.3	0.5	0.218	0.499
21	0.3	0.6	0.212	0.567
22	0.4	0.1	0.417	0.096
23	0.4	0.2	0.366	0.188
24	0.4	0.3	0.381	0.274
25	0.4	0.4	0.342	0.334
26	0.4	0.5	0.344	0.434
27	0.5	0.1	0.507	0.075
28	0.5	0.2	0.465	0.187
29	0.5	0.3	0.474	0.232
30	0.5	0.4	0.430	0.347
31	0.6	0.1	0.607	0.066
32	0.6	0.2	0.580	0.144
33	0.6	0.5	0.555	0.207
34	0.7	0.1	0.686	0.054
35	0.7	0.2	0.732	0.111
36	0.8	0.1	0.790	0.041

References

Banerjee, P. K., & Sinha, B. K. (1979). Generating an event with probability ρ(to the power α), α > 0. *Sankhya; Series B, 41*(3 & 4), 282–285.

Bose, A., & Sinha, B. K. (1985). Unbiased sequential estimation of 1/*p*: Settlement of a conjecture. *Annals of the Institute of Statistical Mathematics, 37*, 455–460.

Girshick, M. A., Mosteller, F., & Savage, L. (1946). Unbiased estimates for certain binomial sampling problems with applications. *The Annals of Mathematical Statistics, 17*, 13–23.

Gupta, M. K. (1967). Unbiased estimate for 1/p. *Annals of the Institute of Statistical Mathematics, Tokyo, 19*, 413–416.
Mohanty, S. G. (1979). *Lattice path counting and applications*. Academic Press.
Rao, C. R. (1952). *Advanced statistical methods in biometrics research*. Wiley.
Sinha, B. K., & Sinha, B. K. (1975). Some problems of unbiased sequential binomial estimation. *Annals of the Institute of Statistical Mathematics, 27*, 245–258.
Sultana, P., Pal, M., & Sinha, B. K. (2023). Exact inference in a tetranomial distribution. *Sequential Analysis, 42*(1), 1–16.
Tarafdar, P. (2016). *Trinomial distributions: Probabilistic and inferential aspects* (Unpublished project report). Indian Institute of Technology, Kanpur, India.
Tarafdar, P., Sultana, P., & Sinha, B. K. (2020). Unbiased sequential binomial/trinomial parameter(s) estimation: An informative review. *International Journal of Statistical Sciences, 20*(2), 57–84.

Chapter 3
Use of Additional Resources in Finite Population Inference

Learning Objectives
Here is yet another fascinating topic the readers will be exposed to. This is in the domain of survey sampling. We fondly hope the readers are familiar with the notions of SRSWR/SRSWOR and related inference—especially in the context of unbiased estimation of a finite population mean or a finite population total. Here we contemplate a situation wherein there is a possibility of availability of additional resources, once the initial phase of choice of a sampling strategy is over. The topic is rich in contents—both theoretically and computationally. The readers will find the exposition thought-provoking and pursuing in various contexts.

3.1 Introduction

In the context of survey sampling, we are all familiar with a popular slogan: More The Merrier! Understandably, one is referring to SRSWR/SRSWOR sampling and further, to estimation of a finite population mean by the corresponding sample mean. At times, one is interested in the finite population total instead of the sample mean. We will not make any distinction between the two and ask for any one of the two interchangeably. However, sampling is hardly restricted to this simple-minded phenomenon. The students and researchers are exposed to a host of sampling designs and unified theory of sampling has grown during the last 70 years or so. Hedayat and Sinha (1991). The notion of 'increasing the sample size' from n to $(n + k)$ for some $1 \leq k \leq (N - n)$, N being the finite population size, has to be understood in proper contexts and its inferential implication should be properly addressed and rigorously studied. This calls for theoretical study in proper contexts of the nature and behavior of sampling strategies which are combinations of sampling designs and the associated estimators.

© The Author(s), under exclusive license to Springer Nature Singapore Pte Ltd. 2024
M. Pal and B. K. Sinha, *Selected Topics in Statistical Inference*,
https://doi.org/10.1007/978-981-97-2592-2_3

We start with the general notion of a sampling design, usually classified as (i) a fixed size sampling design, (ii) a variable size sampling design.

The estimators are generally classified as (i) linear estimators [le], (ii) linear (design-) unbiased estimators [lue], (iii) homogeneous linear (design-) unbiased estimators [hlue].

Standard text books (Hedayat & Sinha, 1991) discuss all these basic notions. A popular estimator for the finite population total is the celebrated Horvitz-Thompson Estimator [HTE].

Unified theory of finite population inference has gone a long way during the last 70 years and any number of textbooks and monographs are readily available. Usual inferential and decision-theoretic concepts have been rigorously introduced and studied.

After a brief review of the basics in Sect. 3.2, in the rest of Sects. 3.3–3.7, we will focus our attention on the study of consequences of increasing the sample size in some sense or the other. This topic has been researched and documented in the published literature. However, usually the researchers are not aware of the intricacies involved and, accordingly, are sold to the popular and simple-minded slogan which holds generally for (SRSWR/SRSWOR, Sample Mean) Strategy !

We believe this study will create renewed interest among students, teachers, and researchers as to the implication of 'increase' in the sample size beyond SRSWR/SRSWOR designs.

3.2 Basics in Finite Population Inference

To start with, let us focus on fixed size (n) sampling designs for a finite labeled population of N units. Otherwise, the sampling designs are quite arbitrary, except that every choice must lead to positive first- and second-order inclusion probabilities, denoted as π_i's and π_{ij}'s. Such a sampling design is also known as a connected sampling design. It is known that for a fixed size (n) sampling design,

$$\text{(i)} \sum_{i} \pi_i = n, \quad \text{(ii)} \sum_{j \neq i} \pi_{ij} = (n-1)\pi_i, \quad i = 1, 2, \ldots, N,$$

$$\text{(iii)} \sum\sum_{i \neq j} \pi_{ij} = n(n-1).$$

There are some other desirable features of sampling designs from inferential perspectives. We refer to Hedayat and Sinha (1991) for details.

The question we pose is the following :

A survey statistician has made a choice of a fixed size (n) sampling design, say $D(N, n)$ with reference to a finite population of N labeled units and is now contemplating on (i) implementation of the design to select a sample $s(n)$ of size n, and (ii) subsequently, drawing n observations $[Y_i | i \in s(n)]$. Once this is done, he is ready to suggest $e(s(n)|Y)$ as a homogeneous linear (design-) unbiased estimator [hlue] of $T(Y)$, the population total of the study feature Y, assumed to be measurable [qualitative / quantitative]. So far this is quite logical and understandable. It is known that there are plenty of choices of such hlues of the type $e(s(n)|Y)$ for the finite population total $T(Y)$ and for any given [non-unicluster] connected sampling design. Hedayat and Sinha (1991) for this and related concepts. Herein above and all through this chapter, we are referring to the only study character Y and it is made amply clear by the use of the notation $e(s(n)|Y)$. When the study feature Y is understood, we may use a simplified notation such as $e(s(n))$. As has been mentioned before, we will not distinguish between the population total and the population mean for inference purpose. Of course, for making comparisons of sampling strategies, we will stick to one of the two.

Of course, these estimators include the celebrated Horvitz-Thompson Estimator HTE which is known to be an hlue of population total/mean whichever we are referring to in a given context.

At this stage, the statistician is told about availability of enhanced resources which may amount to, say a collection of additional data from an 'additional' set of k population units. However, he is well-conversant with the [unified] theory of sampling and so he decides to take recourse to another sampling design $D(N, k)$ without any further specification at this moment. Once this is done, the sampling design $D(N, n)$ will be naturally 'extended' to $D(N, n) \cup D(N, k)$. Note that $e(s(n))$ is the hlue based on a sample $s(n)$ from $D(N, n)$ sampling design. Now the statistician is supposed to suggest another hlue, say $e(s(n, k))$, based on the 'enhanced' sample $s(n, k)$ derived from $D(N, n) \cup D(N, k)$. In order that the use of additional resources becomes a fruitful proposition, the statistician must now ensure that the estimator $e(s(n, k))$, to be proposed judiciously on $D(N, n) \cup D(N, k)$ is at least as good as, and preferably better than, the original estimator $e(s(n))$, irrespective of the choice of the extension design $D(N, k)$, in terms of possessing a smaller variance, both the estimators being hlue in their individual contexts.

Upon considerable thinking, one may note that there is more to it. We will get back to such eventualities later.

3.3 A Basic Result

We first discuss possible interpretation of $D(N, n) \cup D(N, k)$. One interpretation is that the two sampling designs are independently defined over the whole population of N units. We must, however, remember that the two designs are implemented in the sequence $D(N, n)$, followed by $D(N, k)$. In that case, this 'union' refers to a sampling design for which the sample sizes are $(n, n + 1, \ldots, n + k)$. So after implementation of $D(N, n)$ and selection of a sample $s(n)$, we implement $D(N, k)$ independently and arrive at a sample $s(k)$. Then for this 'union' design, the samples are formed by taking 'union' $s(n) \cup s(k)$ of the two component samples. Thus only 'new' units from the population are included in $s(n) \cup s(k)$, besides those contained in $s(n)$. Here data collection efforts are confined to additional units, beyond those in $s(n)$. To fix ideas, let us denote by t the number of additional units observed, beyond the n units in $s(n)$. In other words, we set $|s(n) \cup s(k)| = n + t$. We must note that n units have already been selected and we are then referring to an additional t units, once $D(N, k)$ is implemented independently. Naturally, $t = 0, 1, \ldots, k$. We will assume, WOLG, that $n + k < N$.

Based on the original design $D(N, n)$, the survey statistician has a sampling strategy in mind. By that we are referring to specification of an hlue $e(s(n))$. Estimators such as the Horvitz-Thompson Estimator [HTE], for any arbitrarily chosen connected sampling design $D(N, n)$, Des Raj Estimators and Murthy Estimators for PPSWOR sampling schemes are generally discussed. Hedayat and Sinha (1991) for detailed descriptions of all these sampling designs and related inferential aspects.

We will assume that the readers [students/research scholars/teachers] are familiar with these notions in the context of finite population inference. It is good to know how an hlue is constructed under a very general description of a sampling design. The notion of a connected sampling design is also relevant here. Choice of such a sampling design ensures positive first- and second-order inclusion probabilities of all the population units and of all pairs of population units.

We assume that our choice of the basic sampling design $D(N, n)$ is such that all first- and second-order inclusion probabilities are strictly positive to start with. It is readily seen that this immediately induces similar property of the 'union design', irrespective of the nature of the 'extension' design $D(N, k)$. Moreover, one can verify that the union design is a mixture of designs of the type $D(N, n + t), t = 0, 1, \ldots, k$. Therefore, for any suggested estimator based on the union design, we can study its properties by using the formulae : $E = E_1 E_2$ and $V = V_1 E_2 + E_1 V_2$. This last formula is conventionally and naturally used in the context of a two-stage sampling (Cochran, 1977; Hedayat & Sinha, 1991).

It is time to clarify the nature of formation of an estimator. Naturally, the starting design is $D(N, n)$ and the associated estimator is $e(s(n))$. Next, once the extended design $D(N, n) \cup D(N, k)$ is defined, it amounts to a mixture of $(k + 1)$ designs of the type $D(N, n + t), t = 0, 1, \ldots, k$. For each $t > 0$, we have t additional units, beyond those n units contained in $s(n)$. Therefore, conditional on t, we form an estimator, say $e(s(n + t))$ based on y-values underlying $(n + t)$ distinct units—n of these units being already contained in $s(n)$. To study the properties of the resulting estimator, we make use of the formulae mentioned above.

Recall SRSWR(N, n) sampling design and computation of the explicit expressions for the mean and variance of the sample mean based on distinct units in SRSWR(N, n). We refer to Hedayat and Sinha (1991) for all these basic results and their derivations. More on a comparison of sample mean based on distinct units and sample mean based on all units [including repeates] under SRSWR(N, n) is discussed in Sen and Sinha (1989).

We can pose the inference problem in terms of performance of $D(N, n)$ and $D(N, n) \cup D(N, k)$ and related hlues. We ask : Given $D(N, n)$ and related hlue $e(s(n))$ and the union design $D(N, n) \cup D(N, k)$, can we improve upon $e(s(n))$ by suggesting another hlue, viz. $e(s(n) \cup s(k))$ which has uniformly smaller variance ?

Note that while studying the inferential properties of the proposed hlue, viz. $e(s(n) \cup s(k)|Y)$, the union design is to be understood in terms of the mixture design explained above.

Under the present context, we first establish a positive result toward superiority of the union design [as a mixture design], when both $D(N, n)$ and $D(N, k)$ are independently implemented SRSWOR Designs.

3.4 Case of SRSWOR Designs

To explain the computations, we confine to the particular case : $k = 1$. It is well-known that the strategy [SRSWOR$(N, n+1)$, $\bar{y}(n+1)$] is uniformly superior to the strategy [SRSWOR(N, n), $\bar{y}(n)$] for estimation of the population mean, where $\bar{y}(n)$ is the sample mean based on a sample of size n. Next, note that the union design SRSWOR$(N, n) \cup$ SRSWOR$(N, 1)$ is a 'mixture' sampling design with two components, viz.

$$\text{(i) SRSWOR}(N, n), \quad \text{(ii) SRSWOR}(N, n+1),$$

where the mixing proportions are n/N and $(N - n)/N$ respectively.

It may further be noted that each of these sampling designs is permutation invariant in the population units $(1, 2, \ldots, N)$.

Using the formulae

(i) $E = E_1 E_2$ and (ii) $V = V_1 E_2 + E_1 V_2$ and also using the sample mean in each of the component designs for estimation of the population mean $(\bar{Y})$, we derive

$$\begin{aligned}
(a) E &= E1E2 = \bar{Y}, \\
(b) V &= E1V2 + V1E2 = S^2[(N-n)/Nn] \times [n/N] \\
&\quad + S^2[(N-n-1)/N(n+1)] \times ((N-n)/N] \\
&= S^2[(N-n)/Nn] \times n/(n+1)
\end{aligned}$$

and this last expression in bounded by $S^2[(N-n)/Nn]$ and $S^2[(N-n-1)/N(n+1)]$.

Hence the claim is verified. Note that in the context of SRSWOR, S^2 refers to the population variance with divisor $(N - 1)$.

In the general case, the 'boundary' or 'extreme' designs are SRSWOR(N, n) and SRSWOR$(N, n+k)$ and naturally, the same arguments, i.e., $E = E_1 E_2$ and $V = E_1 V_2 + V_1 E_2$ hold good and we have the same inequality result as above.

However, it would still be interesting to derive an exact expression for $V = E_1 V_2 + V_1 E_2$ underlying the mixture design or the union design SRSWOR(N, n)

∪ SRSWOR$(N, n+k)$ for a general value of $k > 1$. We refer to a standard formula connecting the binomial coefficients.

$$N_{C_k} = n_{C_k} + n_{(k-1)}(N-n)_1 + n_{(k-2)}(N-n)_2 + \cdots + (N-n)_k, \quad k = 0, 1, 2, \ldots$$

With the above representation, an explicit expression for V can now be deduced. We encourage the students and researchers to accomplish this and verify the variance inequalities.

3.5 Case of Arbitrarily Specified $FS(n)$ Connected Design and an Extended Mixture Design

To fix ideas, we start with one example [and one exercise] to illustrate all the operations. The basic sampling design is quite arbitrary except that it is a connected design—implying thereby that all first- and second-order inclusion probabilities are strictly positive.

Example 3.5.1 We choose $N = 6, n = 3, k = 1$ and start with $D(N = 6, n = 3)$ as shown below.

Sample of size three	(123)	(145)	(126)	(246)	(256)	(345)	(346)
Probability	0.10	0.15	0.15	0.15	0.20	0.05	0.20

Note that the extension design of size 1 may or may not be defined over the whole population of 6 units. Accordingly, we discuss two different examples, though both will serve as extensions of the chosen design displayed above.

For a subset selection of the extension design, i.e., of $D(N = 6, k = 1)$, we use $P(1) = 0.3$, $P(4) = 0.7$ and this is applied independently of $D(N = 6, n = 3)$. Naturally, the union design will have 2 different sample sizes, viz. 3 and 4 and, accordingly, it will be a varying size sampling design. We show the union design below after computing the probabilities according to the product law under independence.

Sample of size three	(123)	(145)	(126)	(246)	(345)	(346)	TOTAL
Probability	0.03	0.15	0.045	0.105	0.035	0.140	0.505
Sample of size four	(1234)	(1246)	(1256)	(2456)	(1345)	(1346)	TOTAL
Probability	0.07	0.15	0.06	0.14	0.015	0.06	0.495

The above sample compositions of the extended sampling design are self-explanatory and so are the probability computations. We now start with an hlue of $T(Y)$, say the HTE$(s(n)) = \sum_i [Y_i/\pi_i]$ based on $D(N, n)$ design. It is a routine task to compute mean and variance of the HTE for the above example. We are asking the question : Does there exist an improved hlue of $T(Y)$ based on the union design ?

Lanke (1975) resolved this problem in the afffirmative. His statement is even more powerful. He asserted that the union design can be further extended to a $FS(n+k)$ Design, say $D^*(N, n+k)$ and an hlue of $T(Y)$ can be defined on the extended design by combining the 'old' and 'new' units and this later estimator fares better than the hlue [HTE] based on the original design. We first explain the extension rule.

Note that the extended sampling design $D^*(N=6, n+k=4)$ is defined as a fixed size (4) sampling design over $D(N=6, n=3) \cup D(N=6, k=1)$ by using all the samples of size 4 listed above and it is confined to those samples alone. The probabilities are to be enhanced by using a suitable formula. The formula takes care of the $P(s(n))$ values for all the samples of size $n=3$ under $D(N=6, n=3) \cup D(N=6, k=1)$ listed above. Once done, the total probability under the new design $D^*(N=6, n+k=4)$ will be $0.505 + 0.495 = 1.00$, as it should.

We reproduce the above list of samples of size 4 and the associated probabilities.

Sample of size four	(1234)	(1246)	(1256)	(2456)	(1345)	(1346)	TOTAL
Probability	0.07	0.15	0.06	0.14	0.015	0.06	0.495

Below in the table we display (i) sample(s) of size 3 of the union design, (ii) sample(s) of size 3 [union design] extended to sample(s) of size 4 of the extended design and (iii) merger of these samples with union design-based samples of size 4 to arrive at samples of the extended design $D^*(N=6, n+k=4)$ of $FS(4)$.

Sl. No.	Sample of size 3 under union	Probability under union	Nature of merged sample of size 4	Combined prob. under FS(4) design $P(s(4))$
1	(123)	0.030	(1234)	$0.070 + 0.030 = \mathbf{0.100}$
2*a*	(145)	0.150	(1345)	$0.015 + 0.150 = 0.165^*$
2*b*	(345)	0.035	(1345)	$(addl.)\ 0.035 = 0.035^*$
2	–	–	(1345)	$0.165 + 0.035 = \mathbf{0.200}$
3*a*	(126)	0.045	(1246)	$0.150 + 0.045 = 0.195^*$
3*b*	(246)	0.105	(1246)	$(addl)\ 0.105 = 0.105^*$
3	–	–	(1246)	$0.195 + 0.105 = \mathbf{0.300}$
4	(346)	0.140	(1346)	$0.060 + 0.140 = \mathbf{0.200}$
5	–	–	(1256)	$0.060 + - = \mathbf{0.060}$
6	–	–	(2456)	$0.140 + - = \mathbf{0.140}$
Total	–	–	–	$0.495 + 0.505 = \mathbf{1.000}$

Note that the choice of the $D(N, k)$ design, i.e., $D(6, 1)$ design rendered the possibility of two sample sizes after union, viz. $n=3$ and $n=4$. We have no issues with the samples of size 4 in the union design. Those are listed comfortably along with the respective probabilities, computed as per the multiplication rule. Now we need to 'push' the samples of size 3 in the union design [column 2 above] to 'suitably identified' samples of size 4—among those already available under the union design. We scan through this list one by one.

(i) $(1, 2, 3)$: It can be merged only with $(1, 2, 3, 4)$. That is indicated under Sl. No. 1 in columns (4) and (5).
(ii) $(1, 4, 5)$ and $(3, 4, 5)$: These two can be merged only with $(1, 3, 4, 5)$. These are shown under Sl. No. 2a, 2b, and 2.
(iii) $(1, 2, 6)$ and $(1, 4, 6)$: These two can be merged only with $(1, 2, 4, 6)$. These are shown under Sl. No. 3a, 3b, and 3.
(iv) $(3, 4, 6)$: It can only be merged with $(1, 3, 4, 6)$. That is indicated under Sl. No. 4 in columns (4) and (5).

There is an interesting implication of this result. First, extra funding is available, and hence one or more $k \geq 1$ additional and distinct population unit(s) can be selected and the corresponding Y-value(s) obtained. So, if the union design is chosen in a way that the two components $D(N = 6, n = 3)$ and $D(N = 6, k = 1)$ act independently, we may end up effectively in a resulting sample of size $n = 3$, instead of a sample of size $n + k = 3 + 1 = 4$ for some sample compositions. The above procedure, due to Lanke (1975), suggests that the extended sampling design is indeed a $FS(n + k = 4)$ design, and hence it keeps the spirit of utilization of additional resources in action. So even a deliberate action of the sampler can be rectified by this approach.

On the other hand, to start with, one can suitably define the union design in a way that it already turns out to be a $FS(n + k)$ design. We will study this aspect later after Remark 2.

Now we need to focus on the estimation aspect. Again, to start with, we have an hlue $e(s(n))$ in mind, based on the $D(N, n)$ sampling design. This could be, for example, the HTE. We claim that it is possible to suggest a better estimator based on D^* design [which is a $FS(4)$ design] in the sense of uniformly smaller variance than that of the hlue $e(s(n))$ we start with.

The proof is based on two steps, and it utilizes the union design $D(N = 6, n = 3)$ $\cup\, D(N = 6, k = 1)$, which is composed of samples of sizes $n = 3$ and $n = 4$.

At the first step, we demonstrate that the extended design can be viewed as

$$P(s(n+k)) = \sum_{s(n)\ \text{subset of}\ s(n+k)} P[s(n)|D(N,n)]Q(s(n+k) - s(n))$$

for some conditional probability $Q(.|.)$ defined for every $s(n)$ for which $P(s(n)|D(N, n)) > 0$.

We showcase this phenomenon here for the above example.

Sl. No.	Sample of size 4 under extended design	Probability	Nature of $Q(.\|.)$
1	(1234)	0.100	$Q(4\|(123)) = 1$
2	(1345)	0.200	$Q(3\|(145)) = 1$; $Q(1\|(345)) = 1$
3	(1246)	0.300	$Q(4\|(126)) = 1$, $Q(1\|(246)) = 1$
4	(1346)	0.200	$Q(1\|(346)) = 1$
5	(1256)	0.060	$Q(1\|(256)) = 0.3$, $Q(4\|(256)) = 0.7$
6	(2456)	0.140	$Q(1\|(256)) = 0.3$, $Q(4\|(256)) = 0.7$

It is interesting to note that in the original FS$(n = 3)$ design, there are 7 samples. Except for the sample (256), all others are extended by a single-point design and (256) is extended by a 2-point design. This phenomenon of conditional distribution is generally true.

We are now in a position to recommend an improved estimator, starting with an estimator $e(s(n))$ based on $D(N, n)$. This we do by referring to a result due to Lanke (1975).

We assume that an hlue is already available based on $D(N = 6, n = 3)$ design. We use and improve upon the same. The formation of the improved estimator is based on Rao-Blackwellization principle.

Sl. No.	Sample $s(n + k)$ under extended design	Probability	$e(s(n + k))$
1	(1234)	0.100	$e(1234) = e(123)$
2	(1345)	0.200	$e(1345) = [e(145)P(145) + e(345)P(345)]/P(s(1345))$
3	(1246)	0.300	$e(1246) = [e(126)P(126) + e(246)P(246)]/P(s(1246))$
4	(1346)	0.200	$e(1346) = e(346)$
5	(1256)	0.060	$e(1256) = e(256)$
6	(2456)	0.140	$e(2456) = e(256)$

It trivially follows that

$$\text{(i) } E[e(s(n + k))] = \sum e(s(n + k))P(s(n + k))$$

$$= \sum e(s(n))P(s(n)) = E[e(s(n))] = T(Y).$$

$$\text{(ii) } E[e^2(s(n + k))] = \sum\nolimits_1 e^2(s(n + k))P(s(n + k))$$

$$= \sum\nolimits_1 e^2(s(n))P(s(n)) + [\sum\nolimits_2 e(s(n))P(s(n))]^2/P(s(n + k)).$$

In the above, first $\sum_1$ refers to terms arising out of single samples, i.e., Sl. No. 1, 4, 5, 6 and $\sum_2$ refers to terms arising out of mixtures of samples, i.e., Sl. No. 2, 3. Next note that in view of C-S inequality, the $\sum_2$ is dominated by

$$\left[\sum\nolimits_{2} e^2(s(n))P(s(n))\right],$$

and this encompasses 4 samples listed in 2, 3. Combining the two $\sum$s, we assert the claim that the strategy $[D(s(n+k)), e(s(n+k))]$ improves over the starting strategy $D(s(n)), e(s(n))$.

Remark 3.5.1 In the above, we started with an extension design $D(N = 6, k = 1)$ independently of $D(N = 6, n = 3)$. The extension design was taken as a 2-point design. Here we provide an instructional exercise to the students and research scholars. Take the extension design, still with $k = 1$, as

$$P(1) = 0.30, \quad P(2) = 0.10, \quad P(3) = 0.05, \quad P(4) = 0.15,$$
$$P(5) = 0.25, \quad P(6) = 0.15.$$

Work out an improved strategy, starting with any hlue based on the $D(N = 6, n = 3)$ design already introduced and studied. The improvement arises out of enhancement of the FS$(n = 3)$ to FS$(n + k = 4)$.

Remark 3.5.2 So far we discussed the nature of extension of a given sampling design by using an independent sampling design. That generally results in varying size sampling designs and, therefore, use of additional resources is not necessarily reflected in the data analysis.

Now we propose a different sampling scheme which promotes the use of a $FS(n + k)$.

Suppose the survey statistician has already implemented the sampling design $D(N, n)$ and arrived at a sample $s(n)$. At this stage, he is asked to adopt a 'conditional' sampling design $D(N, k)|s(n)$, defined on the 'complimentary' part of $s(n)$ which forces him to make a distinct choice of additional k units. Therefore, the union design $D(N, n, s(n)) \cup D((N, k)|s(n))$ turns out to be a $FS(n + k)$ design. It would be an interesting exercise to demonstrate that irrespective of the choice of the extension design $D((N, k)|s(n))$, an initially chosen sampling strategy $[D(N, n), e(s(n))]$ can be uniformly improved upon by another strategy based on the extended design, which is now necessarily a $FS(n + k)$ sampling design. We leave the verification to interested readers and researchers.

There are still so many issues involved regarding the use of additional resources. We will turn to use of Lanke's formula at the end.

Remark 3.5.3 In as early as 1967, Prabhu-Ajgaonkar discussed the possibility of an hlue based on a sample of size n to outperform an hlue based on an extended design of size $n + 1$—the two estimators belonging to the same class of hlues. The sampling design was chosen to be the Midzuno sampling scheme for a sample of size $n = 2$ and it was extended to the generalized Midzuno sampling scheme for a sample of size $n = 3$. However, he actually worked out the case of $n = 1$ against $n = 2$ and that was not at all appealing. Our attempt to work for $n = 2$ did not go through with the set-up adopted by him.

Starting with an arbitrarily specified initial FS(N, n) design and extending it to an arbitrary FS($N, n+k$) design by increasing the sample size from n to $n+k < N$, confining to the use of the HTE in both the designs, one can not uniformly improve over the HTE based on the original design. A quick and tricky proof goes like this. Let $(\pi_i(n)|FS(N, n); i = 1, 2, \ldots, N)$ denote the first-order inclusion probabilities based on the original design so that $\sum(\pi_i(n)|\text{FS}(N, n)) = n$. Set $Y_i = K\pi_i(n)/n, i = 1, 2, \ldots, N$ where K is an arbitrary positive constant. Then $\text{HTE}(s(n)) = \sum_{i \in s(n)} Y_i/\pi_i(n) = K$ for every $s(n)$ with $P(s(n)) > 0$. Hence, $V(\text{HTE}) = 0$ at the stated values of the Y_i's. In the same vein,

$$\text{HTE}(s(n+k)) = \sum_{i \in s(n+k)} Y_i/\pi_i(n+k) = [K/n]\left[\sum_{i \in s(n+k)} \pi_i(n)/\pi_i(n+k)\right].$$

Therefore, unless $\pi_i(n)/\pi_i(n+k)$ is the same for all $i = 1, 2, \ldots, N$, the second estimator has a strictly positive variance at the same stated values of the Y's.

Remark 3.5.4 One might wonder if this result is in contrast with the claim made in the previous section as to the superiority of the design based on the extended design. We must note that for every estimator $e(s(n))$ based on the initial FS(N, n) design, there is uniformly improved estimator $e(s(n+k))$ based on the extended FS($N, n+k$) design. However, the point is that whereas $e(s(n))$ is an HTE to start with, the other design $e(s(n+k))$ is not necessarily an HTE. It is still there for sure! We remind the reader that the formation of the improved estimator is based on Lanke's principle.

3.6 Further Issues Involving the Use of Additional Resources

Once more, we start with the set-up of PPS sampling with N population units and the normed size measures $p_1, p_2, \ldots, p_N; 0 < p_i < 1, i = 1, 2, \ldots, N; \sum p_i = 1$. For completeness, we state Midzuno/generalized Midzuno sampling scheme below in general terms for a sample of size $n = 3$.

$$P(i) = p_i, \quad i = 1, 2, \ldots, N; \quad P(j|i) = 1/(N-1); \quad j \neq i;$$
$$P(k|(i, j)) = 1/(N-2); \quad k \neq j, \neq i.$$

In effect, therefore, we use PPS for the first unit and then take recourse to SRSWOR for the remaining 2 units out of the remaining $N-1$ units, after first unit selection is over. As we see, the selected units are denoted by the ordered triplet (i, j, k) of distinct units. In this context, Des Raj estimator is computed as follows:

$$t_1 = y_i/p_i; \quad t_2 = y_i + (N-1)y_j; \quad t_3 = y_i + y_j + (N-2)y_k$$

and, finally,

$$\bar{t} = [t_1 + t_2 + t_3]/3, \quad V[\bar{t}] = [V(t_1) + V(t_2) + V(t_3)]/9.$$

It is argued that the t_i's are pairwise uncorrelated. In the same vein, for a sample of size $n = 2$, resulting in the ordered sample (i, j), we end up using the estimators t_1 and t_2 and hence the average

$$\bar{t} = [t_1 + t_2]/2; \quad V[\bar{t}] = [V(t_1) + V(t_2)]/4.$$

To distinguish between the two variance expressions for sample sizes $n = 2$ and $n = 3$, we will use the notations V_2 and V_3. A counterexample to the variance comparison amounts to establishing that $V_2 < V_3$, i.e., $5(V(t_1) + V(t_2)) < 4V(t_3))$ since the t_i's are known to be uncorrelated pairwise. As was mentioned earlier, we failed in carrying out that exercise.

We now consider Symmetrized Des Raj Estimator [also known as Murthy's Estimator] for a sample of size $n = 2$. These are fairly known results and we refer to Hedayat and Sinha (1991) for details.

We go through the following steps.

1. Des Raj Estimator for an ordered sample (i, j) of size 2 along with $p(i, j)$:

$$\text{DRE}(i, j) = [(y_i/p_i) + (y_i + (N-1)y_j)]/2,$$
$$p(i, j) = p_i/(N-1), 1 \le i \ne j \le N.$$

2. Likewise, Des Raj Estimator for an ordered sample (j, i) of size 2 along with $p(j, i)$:

$$\text{DRE}(j, i) = [(y_j/p_j) + (y_j + (N-1)y_i)]/2,$$
$$p(j, i) = p_j/(N-1), 1 \le i \ne j \le N.$$

3. Both the estimators above are ordered estimators. Murthy's unordeing or symmetrization principle [Rao-Blackwellization] produces an improved estimator, known as symmetrized Des Raj Estimator as given below.

$$\text{SDRE}(i, j) = \frac{[(y_i/p_i) + (y_i + (N-1)y_j)] \times p_i/(N-1) + [(y_j/p_j) + (y_j + (N-1)y_i)] \times p_j/(N-1)]}{2 \times [p_i/(N-1) + p_j/(N-1)]}$$

We also note that $\tilde{p}(i, j) = p(i, j) + p(j, i) = (p_i + p_j)/(N-1)$ stands for $P(i, j) + P(j, i)$ which is the probability under symmetrization of the unordered pair $(i, j), 1 \le i < j \le N$.

In the computations to follow, we use this ratio as $N(i, j)/D(i, j)$. We may still simplify the ratio as :

$$N(i, j) = (y_i + y_j) + (y_i p_i + y_j p_j) + (N - 1)(y_i p_j + y_j p_i);$$
$$D(i, j) = 2(p_i + p_j).$$

4. As against SDRE(i, j), we may as well use the HTE(i, j) [which is trivially a symmetrized estimator] given by

$$\text{HTE}(i, j) = y_i/\pi_i + y_j/\pi_j, \pi_i = p_i + (1 - p_i)/(N - 1),$$
$$\pi_j = p_j + (1 - p_j)/(N - 1).$$

It follows that $\sum_i \pi_i = 2(= n)$ since it is basically a $FS(n = 2)$ sampling design. Moreover, joint inclusion probability for the pair of units (i, j) is given by

$$\pi_(i, j) = (p_i + p_j)/(N - 1), 1 \le i \neq j \le N.$$

It is readily verified that $\sum_{j(\neq i)} \pi_{ij} = \pi_i[= (n - 1)\pi_i]$. We propose to compare these two estimators [SDRE and HTE] for a sample of size 2 and then again for a sample of size 3, using Generalized Midzuno Scheme for the third unit. It amounts to using SRSWOR for drawing the third unit from the remaining $(N - 2)$ population units.

We take up the following example.

Example 3.6.1 We choose $N = 5, n = 2, k = 1$ and adopt Midzuno sampling scheme for sample size $n = 2$ and extend it to $n = 3$, i.e., by taking $k = 1$ and applying SRSWOR as usual.

Population Units	(1)	(2)	(3)	(4)	(5)	TOTAL
p-values	0.10	0.15	0.20	0.25	0.30	1.00
Y-values	1.25	2.10	2.40	3.25	3.60	12.60

Below we show the detailed computations for SDRE.

$$N(i, j) = (y_i + y_j) + (y_i p_i + y_j p_j) + 4(y_i p_j + y_j p_i);\ D(i, j) = 2(p_i + p_j).$$

$$N(1, 2) = 3.35 + (0.125 + 0.315) + 4 \times [0.1875 + 0.2100]$$
$$= 0.3975 = 3.35 + 0.44 + 1.59 = 5.38,\ D(1, 2) = 0.50$$

$$N(1, 3) = 3.65 + (0.125 + 0.480) + 4 \times (0.25 + 0.24) = 0.49$$
$$= 3.65 + 0.605 + 1.96 = 6.215,\ D(1, 3) = 0.60$$

$$N(1, 4) = 4.50 + [0.125 + 0.8125] + 4 \times [0.3125 + 0.325] = 0.6375$$

$$= 4.500 + 0.9375 + 2.550 = 7.9875,\ D(1, 4) = 0.70$$

$$N(1,5) = 4.85 + (0.125 + 1.080) + 4 \times (0.375 + 0.360 =)0.735$$

$$= 4.850 + 1.205 + 2.940 = 8.995,\ D(1,5) = 0.80$$

$$N(2,3) = 4.5 + [0.315 + 0.48] + 4x[0.420 + 0.360] = 0.780$$

$$= 4.50 + 0.795 + 3.12 = 8.415,\ D(2,3) = 0.70$$

$$N(2,4) = 5.35 + [0.315 + .8125] + 4 \times [0.525 + 0.4875] = 1.0125$$

$$= 5.35 + 1.1275 + 4.05 = 10.5275,\ D(2,4) = 0.80$$

$$\begin{aligned} N(2,5) &= 5.70 + [0.315 + 1.08] + 4 \times [0.63 + 0.54] \\ &= 1.17 = 5.7 + 1.395 + 4.68 = 11.775,\ D(2,5) = 0.90 \end{aligned}$$

$$\begin{aligned} N(3,4) &= 5.65 + [0.48 + 0.8125] + 4 \times [0.60 + 0.65] = 1.25 \\ &= 5.65 + 1.2925 + 5.00 = 11.9425,\ D(3,4) = 0.90 \end{aligned}$$

$$\begin{aligned} N(3,5) &= 6.00 + [0.48 + 1.08] + 4 \times [0.72 + 0.72] = 1.44 \\ &= 6.00 + 1.56 + 5.76 = 13.32,\ D(3,5) = 1.00 \end{aligned}$$

$$\begin{aligned} N(4,5) &= 6.85 + [0.8125 + 1.08] + 4 \times [0.975 + 0.90] = 1.875 \\ &= 6.85 + 1.8925 + 7.50 \\ &= 16.2425,\ D(4,5) = 1.10 \end{aligned}$$

Before computing $E[\text{SDRE}]$, we need to evaluate an expression for $p(i, j)$ in its *symmetrized* form. Naturally, that amounts to evaluation of $\tilde{p}(i, j) = p(i, j) + p(j, i) = (p_i + p_j)/(N - 1)$.

For the case of $N = 5$ and recalling the expression for $D(i, j)$, we find that $\tilde{p}(i, j) = D(i, j)/8$. Therefore, $E(\text{SDRE}(i, j)) = \sum\sum_{i<j} N(i, j)/8$. We now proceed to carry out the computations.

$$\begin{aligned} E[\text{SDRE}] &= \sum_{i<j} N(i, j)/8 = [5.38 + 6.215 + 7.9875 + 8.995] \\ &\quad + [8.515 + 10.5275 + 11.775] + [11.9425 + 13.32] + [15.7425]/8 \\ &= 12.60\ \text{[This is an exact result]}. \end{aligned}$$

Next, we compute $V(\text{SDRE}(i, j))$. Instead, for variance comparison purpose, we may as well compute $E[\text{SDRE}(i, j)]^2$ which amounts to computing $\sum_{i<j}[N^2(i, j)/8] \times D(i, j)$.

We obtain the numerical value of the above expression as

$$[7.2361 + 8.0471 + 11.3929 + 12.6422] + [12.6450 + 17.3169 + 19.2570] + [19.8088 + 22.1778] + [29.9148] = 39.3183 + 49.2189 + 41.9866 + 29.9148 = 160.4336.$$

This leads to the numerical value of the variance of the SDRE as $V(\text{SDRE}) = 1.6735$.

Next, we proceed to carry out necessary computations for the HTE. The basic ingredients are first computed below.

p-values (p_i) : 0.10 0.15 0.20 0.25 0.30 Total 1.00

Y-values (y_i) : 1.25 2.10 2.40 3.25 3.60 Total 12.60

$\pi_i = (1 + 3p_i)/4$: 0.325 0.3625 0.40 0.4375 0.475 Total 2 [*Exact Result*]

y_i/π_i values : 3.8461/2 5.7931 6.00 7.4286 7.5789 [upto 4-places of decimals]

Sl. No.	Sample (i, j)	$\tilde{p}(i, j)$ $(p_i + p_j)/4$	HTE(i, j) $y_i/\pi_i + y_j/\pi_j$	HTE$(i, j) \times \tilde{p}(i, j)$
1	(1, 2)	0.25/4	9.6392	0.6024
2	(1, 3)	0.30/4	9.8461	0.7384
3	(1, 4)	0.35/4	11.2747	0.9865
4	(1, 5)	0.4/4	11.4250	1.1425
Total	–	–	–	3.4700
5	(2, 3)	0.35/4	11.7931	1.0319
6	(2, 4)	0.4/4	13.2217	1.3222
7	(2, 5)	0.45/4	13.3720	1.5043
Total	–	–	–	3.8584
8	(3, 4)	0.45/4	13.4286	1.5107
9	(3, 5)	0.50/4	13.5789	1.6974
Total	–	–	–	3.2081
10	(4, 5)	0.55/4	15.0075	2.0635
Grand	Total	–	–	12.6000

Next step would be, as before, to compute $V(\text{HTE}(i, j))$. Instead, for variance comparison purpose, we will compute $E[\text{HTE}(i, j)]^2$, i.e., HTE$(i, j) \times \tilde{p}(i, j) \times$

HTE(i, j). This is done in the groups as in the above.

$$\text{Group I: } 37.2526;\ \text{II: } 49.7656;\ \text{III: } 43.3354;\ \text{IV: } 30.9680;\ \text{Total: } 161.3216$$

This yields : $V(\text{HTE}) = 2.5616$. On the other hand, $V(\text{SDRE}) = 1.6735$.

This is an interesting result. This suggests that for these population values, SDRE has an edge over the HTE in terms of smaller variance. On the other hand, it is well-known that the HTE is an admissible estimator within the class of hlues of the population total (Hedayat and Sinha 1991). Therefore, these two estimators are competitors against each other.

We now proceed toward developing the corresponding estimators based on a sample of size 3, following Midzuno generalized sampling scheme which amounts to drawing a third unit from the remaining $(N-2)$ population units by SRS. We will generalize SDRE of size 2 to an SDRE of size 3 and the HTE of size 2 to that of size 3.

We recall the framework of the reference population.

Popl. Units	:	1	2	3	4	5	Total
p-values (p_i)	:	0.10	0.15	0.20	0.25	0.30	1.00
Y-values (y_i)	:	1.25	2.10	2.40	3.25	3.60	12.60

Also we rework on the theoretical framework for $n = 3$ in terms of the ordered estimator (i, j, k) first, before doing the symmetrization.

1. Des Raj Estimator for an ordered sample (i, j, k) of size 3 along with $p(i, j, k)$:

$$\text{DRE}(i, j, k) = [(y_i/p_i) + (y_i + (N-1)y_j) + (y_i + y_j + (N-2)y_k]/3,$$

$$p(i, j, k) = p_i/(N-1)(N-2), 1 \leq i \neq j \neq k \leq N.$$

2. Likewise, we need to evaluate expressions for the DRE and $p(\ldots)$ for 5 other permutaions of (i, j, k). Then we can symmetrize and arrive at SDRE for a sample of size 3.
3. Not to obscure the essential steps of reasoning with heavy algebraic expressions, we straightaway go to the computations. Note that $N = 5$ and $n = 3$.

For the entries in the table below, we provide the following explanations:

(1) Serial No., (2) sample (i, j, k), (3) $p(i, j, k) = p_i/12$,

(4) $\text{DRE}(i, j, k) = [(y_i/p_i) + (y_i + 4y_j) + (y_i + y_j + 3y_k)]/3$,
(5) $= (3) \times (4)$, (6) $= (4) \times (5)$

(1)	(2)	(3)	(4)	(5)	(6)
1	(1, 2, 3)	0.1/12	[12.50 + 9.65 + 10.55]/3 = 32.70/3	3.270/36	106.929/108
2	(1, 3, 2)	0.1/12	[12.50 + 10.85 + 9.95]/3 = 33.30/3	3.330/36	110.889/108
3	(2, 1, 3)	0.15/12	[14.00 + 7.10 + 10.55]/3 = 31.65/3	4.7475/36	150.25834/108
4	(2, 3, 1)	0.15/12	[14.00 + 11.70 + 8.25]/3 = 33.95/3	5.0925/36	172.8904/108
5	(3, 1, 2)	0.20/12	[12.00 + 7.40 + 9.95]/3 = 29.35/3	5.8700/36	172.2845/108
6	(3, 2, 1)	0.20/12	[12.00 + 10.80 + 8.25]/3 = 31.05/3	6.2100/36	192.8205/108
	TOTAL	0.90/12	–	28.52/36	906.0717/108
7	(1, 2, 4)	0.1/12	[12.50 + 9.65 + 13.10]/3 = 35.25/3	3.525/36	124.2562/108
8	(1, 4, 2)	0.1/12	[12.50 + 14.25 + 10.80]/3 = 37.55/3	3.755/36	141.0002/108
9	(2, 1, 4)	0.15/12	[14.00 + 7.10 + 13.10]/3 = 34.20/3	5.130/36	175.446/108
10	(2, 4, 1)	0.15/12	[14.00 + 15.10 + 9.10]/3 = 38.20/3	5.730/36	218.886/108
11	(4, 1, 2)	0.25/12	[13.00 + 8.25 + 10.80]/3 = 32.05/3	8.0125/36	256.8006/108
12	(4, 2, 1)	0.25/12	[13.00 + 11.65 + 9.10]/3 = 33.75/3	8.4375/36	284.7656/108
	TOTAL	–	–	34.59/36	1201.1547/108
13	(1, 2, 5)	0.1/12	[12.50 + 9.65 + 14.15]/3 = 36.30/3	3.630/36	131.769/108
14	(1, 5, 2)	0.1/12	[12.50 + 15.65 + 11.15]/3 = 39.30/3	3.930/36	154.449/108
15	(2, 1, 5)	0.15/12	[14.00 + 7.10 + 14.15]/3 = 35.25/3	5.2875/36	186.38734/108
16	(2, 5, 1)	0.15/12	[14.00 + 16.50 + 9.45]/3 = 39.95/3	5.9925/36	239.4004/108
17	(5, 1, 2)	0.30/12	[12.00 + 8.60 + 11.15]/3 = 31.75/3	9.525/36	302.4187/108
18	(5, 2, 1)	0.30/12	[12.00 + 12.00 + 9.45]/3 = 33.45/3	10.035/36	335.6707/108
	TOTAL	–	–	38.40/36	1350.0952/108

(1)	(2)	(3)	(4)	(5)	(6)
19	(1, 3, 4)	0.1/12	[12.50 + 10.85 + 13.40]/3 = 36.75/3	3.675/36	135.05625/108
20	(1, 4, 3)	0.1/12	[12.50 + 14.25 + 11.70]/3 = 38.45/3	3.845/36	147.84025/108
21	(3, 1, 4)	0.20/12	[12.00 + 7.40 + 13.40]/3 = 32.80/3	6.560/36	215.168/108
22	(3, 4, 1)	0.20/12	[12.00 + 15.40 + 9.40]/3 = 36.80/3	7.360/36	270.848/108
23	(4, 1, 3)	0.25/12	[13.00 + 8.25 + 11.70]/3 = 32.95/3	8.2375/36	271.425625/108
24	(4, 3, 1)	0.25/12	[13.00 + 12.85 + 9.40]/3 = 35.25/3	8.8125	310.640625/108
	TOTAL	–	–	38.49/36	1350.97875/108
25	(1, 3, 5)	0.1/12	[12.50 + 10.85 + 14.45]/3 = 37.80/3	3.780/36	142.884/108
26	(1, 5, 3)	0.1/12	[12.50 + 15.65 + 12.05]/3 = 40.20/3	4.020/36	161.604/108
27	(3, 1, 5)	0.20/12	[12.00 + 7.40 + 14.45]/3 = 33.85/3	6.770/36	229.1645/108
28	(3, 5, 1)	0.20/12	[12.00 + 16.80 + 9.75]/3 = 38.55/3	7.710/36	297.2205/108
29	(5, 1, 3)	0.30/12	[12.00 + 8.60 + 12.05]/3 = 32.65/3	9.795/36	319.80675/108
30	(5, 3, 1)	0.30/12	[12.00 + 13.20 + 9.75]/3 = 34.95/3	10.485/36	366.45075/108
	TOTAL	–	–	42.56/36	1517.1305/108
31	(1, 4, 5)	0.1/12	[12.50 + 14.25 + 15.30]/3 = 42.05/3	4.205/36	176.82025/108
32	(1, 5, 4)	0.1/12	[12.50 + 15.65 + 14.60]/3 = 42.75/3	4.275/36	182.75625/108
33	(4, 1, 5)	0.25/12	[13.00 + 8.25 + 15.30]/3 = 36.55/3	9.1375/36	333.975625/108
34	(4, 5, 1)	0.25/12	[13.00 + 17.65 + 10.60]/3 = 41.25/3	10.3125/36	425.390625/108
35	(5, 1, 4)	0.30/12	[12.00 + 8.60 + 14.60]/3 = 35.20/3	10.560/36	371.712/108
36	(5, 4, 1)	0.30/12	[12.00 + 16.60 + 10.60]/3 = 39.20/3	11.760/36	460.992/108
	TOTAL	–	–	50.25/36	1951.64675/108

(1)	(2)	(3)	(4)	(5)	(6)
37	(2, 3, 4)	0.15/12	[14.00 + 11.70 + 14.25]/3 = 39.95/3	5.9925/36	239.400375/108
38	(2, 4, 3)	0.15/12	[14.00 + 15.10 + 12.55]/3 = 41.65/3	6.2475/36	260.208375/108
39	(3, 2, 4)	0.20/12	[12.00 + 10.80 + 14.25]/3 = 37.05/3	7.410/36	274.5405/108
40	(3, 4, 2)	0.20/12	[12.00 + 15.40 + 11.95]/3 = 39.35/3	7.870/36	309.6845/108
41	(4, 2, 3)	0.25/12	[13.00 + 11.65 + 12.55]/3 = 37.20/3	9.3000/36	345.96/108
42	(4, 3, 2)	0.25/12	[13.00 + 12.85 + 11.95]/3 = 37.80/3	9.450/36	357.21/108
	TOTAL	–	–	46.27/36	1787.00375/180
43	(2, 3, 5)	0.15/12	[14.00 + 11.70 + 15.30]/3 = 41.00/3	6.15/36	252.15/108
44	(2, 5, 3)	0.15/12	[14.00 + 16.50 + 12.90]/3 = 43.40/3	6.51/36	282.534/108
45	(3, 2, 5)	0.20/12	[12.00 + 10.80 + 15.30]/3 = 38.10/3	7.620/36	290.322/108
46	(3, 5, 2)	0.20/12	[12.00 + 16.80 + 12.30]/3 = 41.10/3	8.220/36	337.842/108
47	(5, 2, 3)	0.30/12	[12.00 + 12.00 + 12.90]/3 = 36.90/3	11.070/36	408.483/108
48	(5, 3, 2)	0.30/12	[12.00 + 13.20 + 12.30]/3 = 37.50/3	11.250/36	421.875/108
	TOTAL	–	–	50.82/36	1993.206/108
49	(2, 4, 5)	0.15/12	[14.00 + 15.10 + 16.15]/3 = 45.25/3	6.7875/36	307.134375/108
50	(2, 5, 4)	0.15/12	[14.00 + 16.50 + 15.45]/3 = 45.95/3	6.8925/36	316.710375/108
51	(4, 2, 5)	0.25/12	[13.00 + 11.65 + 16.15]/3 = 40.80/3	10.200/36	416.16/108
52	(4, 5, 2)	0.25/12	[13.00 + 17.65 + 13.15]/3 = 43.80/3	10.950/36	479.61/108
53	(5, 2, 4)	0.30/12	[12.00 + 12.00 + 15.45]/3 = 39.45/3	11.835/36	466.89075/108
54	(5, 4, 2)	0.30/12	[12.00 + 16.60 + 13.15]/3 = 41.75/3	12.525/36	522.91875/108
	TOTAL	–	–	59.19/36	2509.42425/108

(1)	(2)	(3)	(4)	(5)	(6)
55	(3, 4, 5)	0.20/12	[12.00 + 15.40 + 16.45]/3 = 43.85/3	8.770/36	384.5645/108
56	(3, 5, 4)	0.20/12	[12.00 + 16.80 + 15.75]/3 = 44.55/3	8.910/36	396.9405/108
57	(4, 3, 5)	0.25/12	[13.00 + 12.85 + 16.45]/3 = 42.30/3	10.575/36	447.3225/108
58	(4, 5, 3)	0.25/12	[13.00 + 17.65 + 14.05]/3 = 44.70/3	11.175/36	499.5225/108
59	(5, 3, 4)	0.30/12	[12.00 + 13.20 + 15.75]/3 = 40.95/3	12.285/36	503.07075/108
60	(5, 4, 3)	0.30/12	[12.00 + 16.60 + 14.05]/3 = 42.65/3	12.795/36	545.70675/108
	TOTAL	–	–	64.51/36	2777.1275/108
GRAND	TOTAL	–	–	453.6/36 = 12.60	17343.839221/108

Therefore, $E(\text{DRE}) = 453.6/36 = 12.60$—as expected.
Further, $V(\text{DRE}) = 17343.839221/108 - (12.60)^2 = 160.5911 - 158.7600 = 1.8311$.

Next, we turn our attention to the SDRE. In order to compute variance expression for SDRE, the necessary ingredients are already available in the above table. We extract numerical values of the twin pair

$$\text{(i) } \tilde{p}(i, j, k), \text{ (ii) } \sum_{(i,j,k)} p(i, j, k)\text{DRE}(i, j, k), \quad 1 \le i \ne j \ne k \le 5.$$

For every triplet combination, numerical value of (ii) is used to produce SDRE. Note that SDRE is computed by taking the ratio of (ii) and (i). The formula is stated below.

We prepare a table with these 10 ordered triplets. Note that for every ordered triplet $[(i, j, k), 1 \le i < j < k \le 5]$,

$$\text{SDRE}(i, j, k) = \frac{\sum \text{DRE}(i, j, k) \times p(i, j, k)}{\sum p(i, j, k)}.$$

Notionally, it is argued that contribution to $E[\text{SDRE}^2]$ is carried out by terms like $[\sum x_i \times p_i]^2$ and it is exceeded by $[\sum x_i^2 \times p_i] \times [\sum p_i]$ which contributes to $E[\text{DRE}^2]$.

In the table below, we use column (5) to reproduce $\text{SDRE}(i, j, k) \times \tilde{p}(i, j, k)$ and use additional column (6) to show $[\text{SDRE}(i, j, k)]^2 \times \tilde{p}(i, j, k)$ which is *column* (4) × *column* (5).

In essence, therefore, the columns (1) to (5) stand, respectively, for Sl. No., triplet (i, j, k), $\tilde{p}(i, j, k)$, $\text{SDRE}(i, j, k)$, $\text{SDRE}(i, j, k) \times \tilde{p}(i, j, k)$ and finally, $[\text{SDRE}(i, j, k)]^2 \times \tilde{p}(i, j, k)$.

(1)	(2)	(3)	(4)	(5)	(6)
1	(1, 2, 3)	0.90/12	10.5629	28.52/36	8.3682
2	(1, 2, 4)	1.00/12	11.53	34.59/36	11.0784
3	(1, 2, 5)	1.10/12	11.6061	38.40/36	12.3798
4	(1, 3, 4)	1.10/12	11.6636	38.49/36	12.4703
5	(1, 3, 5)	1.20/12	11.8222	42.56/36	13.9765
6	(1, 4, 5)	1.30	12.8846	50.25/36	17.9847
7	(2, 3, 4)	1.20/12	12.8528	46.27/36	16.5194
8	(2, 3, 5)	1.30/12	13.0308	50.82/36	18.3951
9	(2, 4, 5)	1.40/12	14.0929	59.19/36	23.1711
10	(3, 4, 5)	1.50/12	14.3356	64.51/36	25.6886
	Total	1.00	–	453.60/36 = 12.60	160.0321

This yields : $V(\text{SDRE}) = 160.0321 - 158.76 = 1.2721$ and this is smaller than $V(\text{DRE}) = 1.8311$, as expected.

We now proceed to carry out a similar exercise using the HTE for a sample of size 3 for the same set-up of $N = 5, n = 3$ and the generalized Midzuno sampling scheme.

Recall that

$$p\text{-values } (p_i) \quad : 0.10 \quad 0.15 \quad 0.20 \quad 0.25 \quad 0.30 \quad 1.00$$

$$Y\text{-values } (y_i) \quad : 1.25 \quad 2.10 \quad 2.40 \quad 3.25 \quad 3.60 \quad 12.60$$

$$\pi\text{-values } \; : (1 + p_i)/2 : 0.55 \quad 0.575 \quad 0.60 \quad 0.625 \quad 0.65 \quad \text{Total } 3 \; [\mathit{Exact\ Result}]$$

$$y_i/\pi_i \; values : 2.2727 \quad 3.6522 \quad 4.0000 \quad 5.2000 \quad 5.5385$$

Sl. No.	Sample (i, j, k)	$\tilde{p}(i, j, k)$ $(p_i + p_j + p_k)/6$	$HTE(i, j, k)$ $[y_i/\pi_i + y_j/\pi_j + p_k/\pi_k]$	HTE$(i, j, k) \times \tilde{p}(i, j, k)$	
(1)	(2)	(3)	(4)	(5) = (3) × (4)	(6) = (4) × (5)
1	(1, 2, 3)	0.45/6	9.9249	4.466205/6	44.32664/6
2	(1, 2, 4)	0.50/6	11.1249	5.56245/6	61.88170/6
3	(1, 2, 5)	0.55/6	11.4634	6.30487/6	72.27525/6
4	(1, 3, 4)	0.55/6	11.4727	6.309985/6	72.39256/6
5	(1, 3, 5)	0.60/6	11.8112	7.08672/6	83.70267/6
6	(1, 4, 5)	0.65/6	13.0112	8.45728/6	110.03936/6
7	(2, 3, 4)	0.60/6	12.8522	7.71132/6	99.10743/6
8	(2, 3, 5)	0.65/6	13.1907	8.573955/6	113.09647/6
9	(2, 4, 5)	0.70/6	14.3907	10.07349/6	144.96457/6
10	(3, 4, 5)	0.75/6	14.7385	11.053875/6	162.91754/6
Grand	Total	6.00/6 = 1.00	–	75.6002/6 = 12.6000	964.70412/6

Therefore, $V(\text{HTE}) = 964.70412/6 - (12.60)^2 = 160.78402 - 158.76 = 2.02402$.

We can now summarize our findings for this example and for sample sizes 2 and 3 by tabulating the numerical values of the variances of the estimators.

Estimator	$n = 2$	$n = 3$
DRE	–	1.8311
SDRE	1.6735	1.2721
HTE	2.5616	2.02402

Remark 3.6.1 For both the SDRE and the HTE, we find reduction in variance from $n = 2$ to $n = 3$. It also happens that SDRE performs better than the HTE for both the cases.

3.7 Understanding Lanke's Formula

It was suggested that an application of Lanke's formula ensures superiority of the estimator $e(s(n) \cup s(k))$ over the original estimator $e(s(n))$. This has two parts : (a) description of the union sampling design and its extension to $FS(n + k)$ sampling design, and (b) formation of estimators based on samples of size $(n + k)$. We wish to take up the sampling strategies studied in Sect. 3.6 to understand the implications of Lanke's formula.

We start by re-iterating that the extended design can be viewed as

$$P(s(n + k)) = \sum_{s(n) \text{ subset of } s(n+k)} P[s(n)|D(N, n)]Q(s(n + k) - s(n))$$

for some conditional probability $Q(.|.)$ defined for every $s(n)$ for which $P(s(n)|D(N, n)) > 0$.

In the example that we have considered above in Section 3.6, we already have the description of the extension rule, viz. SRSWOR for the third unit. Therefore, the extended design is easily described and formed for all ordered triplets. The following derivations are standard. We write LSDR$(1, 2, 3)$ for Lanke-based symmetric triplet arising out of the symmetric pairs $(1, 2)$, $(1, 3)$, $(2, 3)$. Likewise, we write LSE$(1, 2, 3)$) for the Lanke-based estimator based on the SDREs defined on the symmetric pairs $(1, 2)$, $(1, 3)$, $(2, 3)$.

$$\begin{aligned}
P[\text{LSE}(1,2,3)] &= [P(\text{SDRE}(1,2))] \times [P((3)|\text{SDRE}(1,2))] \\
&\quad + [P(\text{SDRE}(1,3))] \times [P((2)|\text{SDRE}(1,3))] \\
&\quad + [P(\text{SDRE}(2,3))] \times [P((1)|\text{SDRE}(2,3))] \\
&= \tilde{p}(1,2)/3 + \tilde{p}(1,3)/3 + \tilde{p}(2,3)/3 \\
&= [\tilde{p}(1,2) + \tilde{p}(1,3) + \tilde{p}(2,3)]/3.
\end{aligned}$$

Likewise,

$$\begin{aligned}
&E(\text{LSE}(123)) \\
&= \frac{[\text{SDRE}(1,2) \times \tilde{p}(1,2) + \text{SDRE}(1,3) \times \tilde{p}(1,3) + \text{SDRE}(2,3) \times \tilde{p}(2,3)]}{[\tilde{p}(1,2) + \tilde{p}(1,3) + \tilde{p}(2,3)]}.
\end{aligned}$$

Recall that

$$\begin{aligned}
\text{SDRE}(i,j) &= \frac{[[y_i/p_i + (y_i + 4y_j)] \times p_i/4 + [y_j/p_j + (y_j + 4y_i)] \times p_j/4]/2}{(p_i + p_j)/4} \\
&= [(y_i + y_j) + (y_i p_i + y_j p_j) + 4(y_i p_j + y_j p_i)]/2(p_i + p_j);
\end{aligned}$$

Further, $P(\text{SDRE}(i, j)) = (p_i + p_j)/4$.

We are now in a position to provide an explicit expression for Lanke-based estimator corresponding to SDRE.

We denote the same by LE$(1, 2, 3)$ corresponding to SDRE$(1, 2, 3)$. It turns out that LE$(1, 2, 3)$ = LN$(1, 2, 3)$/LD$(1, 2, 3)$ where

$$\begin{aligned}
\text{LN}(1,2,3) &= E(\text{SDRE}(1,2)) \times P(S(1,2)) + E(\text{SDRE}(1,3)) \times P(S(1,3)) \\
&\quad + E(\text{SDRE}(2,3)) \times P(S(2,3)) \\
&= [(y_1 + y_2) + (y_1 p_1 + y_2 p_2) + 4(y_1 p_2 + y_2 p_1)]/8 \\
&\quad + [(y_1 + y_3) + (y_1 p_1 + y_3 p_3) + 4(y_1 p_3 + y_3 p_1)]/8 \\
&\quad + [(y_2 + y_3) + (y_2 p_2 + y_3 p_3) + 4(y_2 p_3 + y_3 p_2)]/8.
\end{aligned}$$

Further,

$$\begin{aligned}\text{LD}(1,2,3) &= P[\text{LS}(1,2,3)] = [P(S(1,2)) + P(S(1,3)) + P(S(2,3))]/3\\ &= [(p_1+p_2)/4 + (p_1+p_3)/4 + (p_2+p_3)/4]/3 = (p_1+p_2+p_3)/6.\end{aligned}$$

In the following table, we show necessary computations to arrive at a numerical value of V(LSDRE).

Sl. No.	Sym. Sample SS(i, j, k)	P(LSS(i, j, k)) $(p_i + p_j + p_k)/6$	LE(i, j, k) $LN(i, j, k)/LD(i, j, k)$
(1)	(2)	(3)	(4)
1	(1, 2, 3)	0.45/6	20.010/1.8
2	(1, 2, 4)	0.50/6	23.895/2
3	(1, 2, 5)	0.55/6	26.150/2.2
4	(1, 3, 4)	0.55/6	26.145/2.2
5	(1, 3, 5)	0.60/6	28.530/2.4
6	(1, 4, 5)	0.65/6	33.225/2.6
7	(2, 3, 4)	0.60/6	30.885/2.4
8	(2, 3, 5)	0.65/6	33.510/2.6
9	(2, 4, 5)	0.70/6	38.545/2.8
10	(3, 4, 5)	0.75/6	41.505/3
	Total	1.00	302.40/24 = 12.60

Computations:

$$E[\text{LSDRE}] = 302.40/24 = 12.60.$$

$$\begin{aligned}E[\text{LSDRE}^2] &= [400.4001/1.8 + 570.9510/2 + 683.8225/2.2 + 683.5610/2.2\\ &+ 813.9606/2.4 + 1103.9006/2.6 + 953.8832/2.4 + 1122.9201/2.6\\ &+ 1485.7170/2.8 + 1722.6650/3]/24 = 3827.4860/24 = 159.4785.\end{aligned}$$

Consequently, $V(\text{LSDRE}) = 159.4785 - 158.76 = 0.7185$.
We may reproduce the table of variance comparisons once again.

Estimator	$n = 2$	$n = 3$
SDRE	1.6735	1.2721
LSDRE	–	0.7185
HTE	2.5616	2.02402
LHTE	–	1.0812

It transpires that LSDRE for a sample of size 3 fares much better than the SDRE itself as also the HTE. It would be interesting to compare HTE and LHTE and also the two Lanke versions, viz. LSDRE and LHTE. The last entry corresponding to (LHTE, $n = 3$) is derived below with all computational details.

Lanke (1975) considered extending an arbitrary sampling strategy $[D(N, n), e(s(n)|Y)]$ to another strategy $[D(N, m = n + k), e(s(n + k)|Y)]$ via $Q(N - n, k)$ so that $[D(N, m = n + k), e(s(n + k)|Y)]$ is better than $[D(N, n), e(s(n)|Y)]$ irrespective of the choice of $Q(N - n, k)$. He proposed the estimator $e(s(n + k)|Y)$ through the relation

$$e(s(n+k)|Y)[P(s(n+k)] = \sum_{s(n)\in s(n+k)} e(s(n)|Y)P(s(n))Q(s(n+k)-s(n)] \tag{3.7.1}$$

Here summation is over all $s(n)$ [subsets of $s(n + k)$].

Further,

$$P[s(n+k)] = \sum_{s(n)\in s(n+k)} P(s(n))Q(s(n+k)-s(n))], \tag{3.7.2}$$

summation being over all $s(n)$ [subsets of $s(n + k)$]

Explicitly, this yields

$$\begin{aligned} P(123) &= P(12)P(3|(12)) + P(13)P(2|(13)) + P(23)P(1|(23)) \\ &= [(p_1 + p_2)/4] \times 1/3 + \cdots = (p_1 + p_2 + p_3)/6 \end{aligned} \tag{3.7.3}$$

Further,

$$\begin{aligned} \mathrm{LE}((123|Y))P(123) &= e((12)|Y) \times P(12)P(3|(12)) + \cdots \\ &= [y_1/\pi_1 + y_2/\pi_2](p_1 + p_2)/12 + \cdots \\ &= y_1/\pi_1[(p_1 + p_2)/12 + (p_1 + p_3)/12] + \cdots \end{aligned} \tag{3.7.4}$$

Likewise,

$$\mathrm{LE}((124|Y))P(124) = [y_1/\pi_1] \times [(p_1 + p_2)/12 + (p_1 + p_4)/12] + \ldots \tag{3.7.5}$$

Next,

$$\mathrm{LE}((125|Y))P(125) = [y_1/\pi_1] \times [(p_1 + p_2)/12 + (p_1 + p_5)/12] + \ldots \tag{3.7.6}$$

$$\mathrm{LE}((134|Y))P(134) = [y_1/\pi_1] \times [(p_1 + p_3)/12 + (p_1 + p_4)/12] + \ldots \tag{3.7.7}$$

$$\mathrm{LE}((135|Y))P(135) = [y_1/\pi_1] \times [(p_1 + p_3)/12 + (p_1 + p_5)/12] + \ldots \tag{3.7.8}$$

$$\text{LE}((145|Y))P(145) = [y_1/\pi_1] \times [(p_1 + p_4)/12 + (p_1 + p_5)/12] + \ldots \quad (3.7.9)$$

This takes care of all the six sample triplets involving the 1st unit.

Before proceeding further, we also note, from (3.7.4)–(3.7.9) that

$$\sum_{(i,j,k)\in[1,2,3,4,5]} P(i,j,k) = \sum_{(i,j,k)\in[1,2,3,4,5],1\le i<j<k\le 5} [(p_i + p_j + p_k)/6]$$
$$= \sum_i p_i = 1, \text{ as it should.}$$

Hence, in the expression for $E[\text{LE}(s(n+k))]$, we derive coeff. of y_1/π_1 as

$$12[p_1/12] + 3[(p_2 + p_3 + p_4 + p_5)]/12 = p_1 + (1 - p_1)/4 = (1 + 3p_1)/4 = \pi_1.$$

We thus demonstrate in this example that the LHTE is indeed an unbiased estimator for $T(Y)$. This is a statement of fact in general terms.

We now proceed to carry out the numerical computations for the example in hand. We will evaluate the value of $V(\text{LHTE})$.

Sl. No. (1)	Sym. Sample $SS(i, j, k)$ (2)	$P(LSS(i, j, k))$ $(p_i + p_j + p_k)/6$ (3)	$P(LSS(i, j, k) \times LE(i, j, k)$ Expressions like (1.1) (4)	$(5) = (4)^2/(3)$
1	$(1, 2, 3)$	0.45/6	0.7909	8.3403
2	$(1, 2, 4)$	0.50/6	0.9704	11.3001
3	$(1, 2, 5)$	0.55/6	1.0831	12.7975
4	$(1, 3, 4)$	0.55/6	1.0786	12.6914
5	$(1, 3, 5)$	0.60/6	1.1928	14.2277
6	$(1, 4, 5)$	0.65/6	1.3975	18.0277
7	$(2, 3, 4)$	0.60/6	1.2883	16.5972
8	$(2, 3, 5)$	0.65/6	1.4112	18.3829
9	$(2, 4, 5)$	0.70/6	1.6300	22.7734
10	$(3, 45)$	0.75/6	1.7572	24.7020
	Total	1.00	12.60	159.8412

Thus, finally, we end up as $V(\text{LHTE}) = 159.8412 - 158.76 = 1.0812$ and this is inserted above in the table of comparisons.

3.8 Understanding Lanke's Formula: All That Glitters!

This revealing example was provided in Sengupta et al. (1987). Starting with a $FS(n)$ sampling design $P(N, n)$ and an admissible hlue $e(s(n))$, we extend the same to $e(s(n + k))$ by implementing an extension formula in terms of $Q(s(k)|s(n))$ for every $s(n)$ with $P(s(n)) > 0$. Accordingly, we adopt Lanke's formula and uniformly improve over $e(s(n))$ by $e(s(n+k))$. Lanke's version has undoubtedly improved over the estimator $e(s(n))$. But does it also assure that this Lanke-based estimator is at least admissible in the class of hlues based on the extended $FS(n+k)$—design? Unfortunately, this Lanke-based estimator may itself turn out to be inadmissible! This is a very illuminating example in a general form involving the HTE. We reproduce the example below.

Example 3.8.1 We start with the set-up:

$$N = 6, n = 3, P(N = 6, n = 3), k = 2, e(s(n)) \\ = \text{HTE}(s(n)) = e(s(n)) \quad \text{for brevity.}$$

where $P(s(n))$ is shown in the table below. We also display the extension rule $Q(s(k)|s(n))$ and $P(s(n+k))$ for the extended design in the same table.

Sl. No.	Sample $s(n)$	$P(s(n))$	Sample $s(k)\|s(n)$	Sample $s(n+k)$	$Q(s(k)\|s(n))$	$P(s(n+k))$
1	(1, 2, 3)	0.1	(4, 5)	(1, 2, 3, 4, 5)	1.00	*
2	(1, 2, 5)	0.2	(3, 6)	(1, 2, 3, 5, 6)	1.00	0.20
3	(2, 4, 6)	0.1	(3, 5)	(2, 3, 4, 5, 6)	1.00	0.10
4	(2, 3, 4)	0.3	(1, 6)	(1, 2, 3, 4, 6)	1.00	0.30
5	(1, 3, 5)	0.3	(4, 6)	(1, 3, 4, 5, 6)	0.30	0.09
	–	–	(2, 4)	(1, 2, 3, 4, 5)	0.7	$*\ 0.10 + 0.21$
	Total	1.00	–	–	–	1.00

In the above table, the extended sample (1, 2, 3, 4, 5) is arrived at from two different samples: (i) (1, 2, 3), followed by (4, 5) and (ii) (1, 3, 5), followed by (2, 4). Accordingly, the contribution to $P(1, 2, 3, 4, 5)$ is computed as $0.1 \times 1.00 + 0.3 \times 0.7 = 0.31$.

Next, in the following table, we display the HTE and also the LHTE.

Sl. No.	Sample $s(n)$	$P(s(n))$	HTE$(s(n))$	$s(n+k)$	$P(s(n+k))$	LHTE
1	(1, 2, 3)	0.1	$e(1, 2, 3)$	(1, 2, 3, 4, 5)	0.31	$\frac{10 \times e(1,2,3) + 21 \times e(1,3,5)}{31}$
2	(1, 2, 5)	0.2	$e(1, 2, 5)$	(1, 2, 3, 4, 6)	0.30	$e(2, 3, 4)$
3	(2, 4, 6)	0.1	$e(2, 4, 6)$	(2, 3, 4, 5, 6)	0.10	$e(2, 4, 6)$
4	(2, 3, 4)	0.3	$e(2, 3, 4)$	(1, 3, 4, 5, 6)	0.09	$e(1, 3, 5)$
5	(1, 3, 5)	0.3	$e(1, 3, 5)$	(1, 2, 3, 5, 6)	0.20	$e(1, 2, 5)$
--	Total	1.00	–	–	1.00	–

It is readily verified that $E(\text{HTE}) = E(\text{LHTE}) = T(Y)$. Further, by an application of C- S inequality on $10 \times e(1, 2, 3) + 21 \times e(1, 3, 5)$, we can deduce that

$$[10 \times e(1, 2, 3) + 21 \times e(1, 3, 5)]^2 \leq [10 \times e(1, 2, 3)^2 + 21 \times e(1, 3, 5)^2] \times 31.$$

This inequality is now used to deduce that $V(\text{LHTE}) < V(\text{HTE})$. Interestingly, we do not have to analyze the structure of the HTE nor we need the specification of the numerical data ! Sengupta et al. (1987) further indicated that one can construct a few more such improved estimators based on the same original design $P(N, n)$ and the same extended design $P(N, n + k)$.

3.9 Conclusion

In this chapter, we have discussed, more or less extensively, the implications of the notion of 'enhanced' sample size when additional resources are made available after selection and implementation of a sampling design in a given context. Suitably defined 'improved' sampling strategies based on enhanced resources are indeed available in the sense of 'extended' sampling designs. But there are issues that are perplexing in nature. We have tried to raise/explain the points of concern toward the effective use of additional resources.

References

Cochran, W. G. (1977). *Sampling Techniques* (3rd ed.). Wiley.

Hedayat, A. S., & Sinha, B. K. (1991). *Design and inference in finite population sampling*. Wiley.

Lanke, J. (1975). Some contributions to theory of survey sampling (Ph.D. Thesis, University of Lund, Sweden)

Prabhu-Ajgaonkar, S. G. (1967). The effect of increasing sample size on the precision of an estimator. *The American Statistician, 21*(4), 26–28.

Sen, P. K., & Sinha, B. K. (1989). On averaging over distinct units in sampling with replacement. *Sankhya Series B, 51*(1989), 65–83.

Sengupta, S., Sinha, B. K., & Pantula, S. G. (1987). Some inferential aspects of finite population sampling with additional resources. *Journal of Statistical Planning and Inference, 16*, 203–211.

Chapter 4
Notion of Sufficiency in Statistical Inference—Theory and Applications

Learning Objectives
In this chapter, we introduce the concept of 'sufficiency' at the foundational level. Most readers are likely to be already familiar with this concept. Our classroom teaching experience is that most students understand this concept rather mechanically, without any depth. We believe that a serious learner will take full advantage of the ways of introducing the concept, as was developed by Blackwell. This is in the domain of parametric inference. Our treatment, for the most part, is likely to be new to the readers. We fondly hope the reader will grasp the concept through a variety of illustrative examples.

4.1 Introduction: Understanding Sufficiency

To start with, we consider a random variable X (discrete or continuous) having a pmf or a pdf, involving just one unknown parameter θ in a continuous domain $\cap\,\Omega$. Given a random sample $\mathbf{x}$ (of assumed size n) from this population, a statistic $t(\mathbf{x})$ is said to be sufficient for the parameter θ, if one can retrieve the whole of $\mathbf{x}$ from $t(\mathbf{x})$ alone! This is rephrased as: 'The conditional distribution of $\mathbf{X}$, given $t(\mathbf{X})$, is independent of θ'. This is usual textbook notion/definition of sufficiency or a sufficient statistic. Factorization Theorem facilitates an explanation of this notion. We will not elaborate on this approach any further.

Instead, we will introduce below the notion of 'Comparison of Statistical Experiments'. We will closely follow Blackwell (1951, 1953).

We have two statistical experiments $\mathscr{E}$ and $\mathscr{F}$—resulting in, respectively, random realizations of $X = (\mathscr{X}, \mathscr{A}, P_\theta, \theta \text{ in } \cap\,\Omega)$ and $Y = (\mathscr{Y}, \mathscr{B}, Q_\theta, \theta \text{ in } \cap\,\Omega)$ having the respective law of distributions $P(.|\theta)$ and $Q(.|\theta)$, involving a common unknown parameter θ in $\cap\,\Omega$. We say that the experiment $\mathscr{E}$ is sufficient for the experiment

© The Author(s), under exclusive license to Springer Nature Singapore Pte Ltd. 2024
M. Pal and B. K. Sinha, *Selected Topics in Statistical Inference*,
https://doi.org/10.1007/978-981-97-2592-2_4

$\mathscr{F}$ if, upon execution of $\mathscr{E}$ and observing X, we can generate an observation on Y, without actually executing the corresponding experiment $\mathscr{F}$!

Analytically, the experiment leading to X is sufficient for the experiment leading to Y if and only if there is a known (i.e., free from the unknown parameter θ) stochastic transformation $\pi(.|.)$ from X to Y such that

$$Q_\theta(B) = \int \pi(B|x)\mathrm{d}P_\theta(x) \tag{4.1.1}$$

for all B in $\mathscr{B}$ and for all θ in $\cap\,\Omega$.

Note that the known stochastic transformation $\pi(.|.)$ helps in generating a random variable, say Z, via the relation (4.1.1) and this Z is distributionally equivalent to Y. It is in this sense that X is sufficient for Y. This becomes more prominent when applied to continuous distributions.

This notion of sufficiency of one statistical experiment for another statistical experiment extends naturally to multivariate distributions and multiparameter situations. Also it extends naturally to sample sizes more than one. Thus, for example, a realization $(x_1, x_2, \ldots, x_n)$ of a random sample of size n from the reference X-population may be used to generate another random sample $(y_1, y_2, \ldots, y_m)$ of size m from a second reference population, say Y-population via a (known) stochastic transformation $\pi(.|.)$ as in (4.1.1) and which does not involve the unknown parameter(s) at all. Then we qualify the statement of sufficiency of the experiment involving $(x_1, x_2, \ldots, x_n)$ from the distribution on X for the experiment involving $(y_1, y_2, \ldots, y_m)$ from the distribution on Y.

4.2 Examples of Sufficiency in Statistical Experiments

Below we take an example to illustrate this phenomenon of sufficiency in statistical experiments.

Example 4.2.1 Reference pmf corresponds to Bernoulli trial with success probability $\theta, 0 < \theta < 1$ in terms of flip of a coin with 'head' as 'success'. Let Y correspond to the description of the output of flipping the coin 5 times—one by one—as and when the flips are made and let X correspond to the count of heads at the end. Then, it is known that

$$P\left[(y_1, y_2, y_3, y_4, y_5)\right] = \Pi_{i=1}^{(i=5)} \theta^{y_i}(1-\theta)^{(1-y_i)};\ y_i = 0, 1;\ i = 1, 2, 3, 4, 5,$$

and, further,

$$P[X = x] = n_{c_x}\theta^x(1-\theta)^{(n-x)};\ x = 0, 1, 2, 3, 4, 5.$$

It is trivial to argue that

$$P[(y_1, y_2, y_3, y_4, y_5)|X = x] = 0; \sum y_i \neq x; = 1/n_{c_x};$$
$$\sum y_i = x; x = 0, 1, 2, 3, 4, 5.$$

This suggests that X is sufficient for Y; in other words, we assert the well-known result that in a fixed number of Bernoulli trials, the number of heads is a sufficient statistic.

If the actual realization is (S, F, F, S, F), then we may as well hide the details and only provide the head-count of total number of successes $t = 2$ to the statistician. The statistician picks up all possible $5_{c_2} = 10$ possibilities with equal chance for each and decides on one of the ten outcomes to produce a sequence of five outcomes, say z. In doing so, he does not need any knowledge of the unknown parameter θ. Moreover, he will generate—analogously—all possible results for any conceivable value of $\sum y_i$. Therefore, Z will be distributionally equivalent to Y.

It may be noted that data generation does not mean 'reproducing' the 'original' data, given X. One has to reproduce the same law for Z as that of Y. In a way, then, Z may be used as a 'proxy' for Y. That will make sense if the marginal distribution of Z coincides with that of Y. That is exactly what is happening in the above. Binomial law for X and uniform distribution of $Z|X = x$ provides the desired unconditional probability law for Z which coincides with that of Y. This is true independent of any knowledge of θ.

In text books, this is stated in the form of a Theorem.

Theorem 4.2.1 *If $(X_1, X_2, \ldots, X_n)$ is a random sample of size n from the reference Bernoulli population with unknown parameter θ, then the conditional distribution of $(X_1, X_2, \ldots, X_n)$, given $\sum_i X_i = T(= 0, 1, \ldots, n)$ is independent of θ and it is discrete uniform distribution over n_{C_T} possible ordered $n - ples$ of $(n - T)$ F's and T S's.*

Here T is known as a sufficient statistic for θ. It has a binomial distribution with parameters (n, θ). Its distribution is complete and T/n serves as UMVUE of θ.

For unbiased estimation of higher positive integral powers of θ, we land into what are known as 'factorial moments'. For positive non-integer powers of θ [such as $\sqrt{\theta}$], or for negative integer/non-integer powers of θ, we need Sequential Bernoulli Sampling Schemes. This topic has been covered in Chap. 2 in considerable details.

We state some further results below involving sufficiency in binomial and Poisson Distributions.

Theorem 4.2.2 *If X and Y are independent Bernoulli trials with parameters θ and $\theta/2$, respectively, then X is sufficient for [or 'more informative' than] Y w.r.t. the unknown parameter θ. Likewise, if X is Binomial (n, θ) and Y is independent Binomial $(m, \theta/2)$, then X is sufficient for Y when θ is unknown provided $n \geq m$.*

Note that the above result breaks down if $n < m$.

Theorem 4.2.3 *If X and Y are independent Poisson variables with parameters $a\theta$ and $b\theta$, respectively, then X is sufficient for [or 'more informative' than] Y w.r.t. the unknown parameter $\theta > 0$ provided $0 < b < a < \infty$, both being known.*

This result is also well-known, and we skip the proof. The proof rests on the fact that one can generate the conditional distribution of $Y|X = x$ based on binomial distribution with parameters $(x, b/a)$ for $x > 0$ and it is degenerate at $y = 0$ when $x = 0$.

Next, we move to univariate normal distributions with unknown mean and known variances.

Theorem 4.2.4 *If X and Y are independent univariate normal variables with parameters $(\mu, a\sigma^2)$ and $(\mu, b\sigma^2)$, respectively, where μ is unknown and $a, b > 0, \sigma^2 > 0$ are all known, then X is sufficient for [or 'more informative' than] Y w.r.t. the unknown parameter μ provided $0 < a < b < \infty$.*

Theorem 4.2.5 *If X and Y are independent univariate normal variables with parameters (μ, σ_1^2) and (μ, σ_2^2), respectively, where μ is unknown and both $\sigma_1^2 > 0$ and $\sigma_2^2 > 0$ are known, then X is sufficient for [or 'more informative' than] Y w.r.t. the unknown parameter μ provided $\sigma_1^2 < \sigma_2^2$.*

The proofs of all the above results rest on data generation on Z, starting with data on X and then showing that Z is distributionally equivalent to Y. We skip the proofs.

In a normal population, when both the mean and the variance are unknown, it is well-known that the sample mean and the sample sum of squares [SS] are jointly sufficient for the population mean and variance. Textbook proof rests on factorization theorem. An eye-opener as to the generation of data, given the sample mean and sample SS, is provided below for the simple case of n = 3.

Theorem 4.2.6 *If X has univariate normal distribution with parameters (μ, σ^2)—both unknown—then the sample mean and sample SS are jointly sufficient for generating the entire body of data and we provide an explicit derivation for $n = 3$.*

Proof Starting with $n = 3$ observations (x_1, x_2, x_3), we compute the sample mean $\bar{x}$ and the sample Sum of Squares $\mathrm{SS}(x)$. Recall that

(i) $\bar{X}$ is normal with mean μ and variance $\sigma^2/n = \sigma^2/3$;
(ii) $\mathrm{SS}(X)/\sigma^2$ has central Chi-square distribution with $n - 1 = 2$ df;
(iii) $\bar{X}$ and $\mathrm{SS}(X)$ are independently distributed;
(iv) $\mathrm{SS}(X) = (X_1 - X_2)^2/2 + (X_1 + X_2 - 2X_3)^2/6$ and the two components are independently distributed, each as $\sigma^2\times$ Chi-square with 1 df.

Therefore, $3(X_1 - X_2)^2/[(X_1 + X_2 - 2X_3)^2]$ has a central F-distribution with $(1, 1)$ df, independent of the parameters.

We now proceed through the following steps to generate a set of 3 observations, say (y_1, y_2, y_3), based on $\bar{x}$ and $\mathrm{SS}(x)$, which are distributionally equivalent to (x_1, x_2, x_3).

Step I. Draw an observation from $F(1, 1)$-distribution. Call it f. *Step II.* Set observed $\mathrm{SS}(x)$ as $\mathrm{ss}(x)$. Compute

(i) $(y_1 - y_2)^2 = 2f\text{ss}(x)/[1+f]$; (ii) $(y_1 + y_2 - 2y_3)^2 = 6\text{ss}(x)/[1+f]$.

From *Step II*, note that $\text{ss}(y) = \text{ss}(x)$. From (i), solve for $|y_1 - y_2|$ and from (ii), solve for $|y_1 + y_2 - 2y_3|$. Sort out all 4 possibilities and use each one to solve for the y_i's—noting that $\bar{y} = \bar{x}$. The y_i's so generated are distributionally equivalent to the x_i's—but for permutation among themselves. There are four possible solutions.

Example 4.2.2 For $n = 3$, we choose $\bar{x} = 3.5$ and $\text{ss}(x) = 5$. In *Step I*, we find $f = u^2/v^2 = 3.28$. Here u, v are random realizations of U, V, being iid $N(0, 1)$ variates. In *Step II*, we compute

(i) $|(y_1 - y_2)| = \sqrt{2f\text{ss}(x)/[1+f]} = 2.7683$,
(ii) $|(y_1 + y_2 - 2y_3)| = \sqrt{6\text{ss}(x)/[1+f]} = 2.6475$.

Further, $(y_1 + y_2 + y_3) = 10.5$. There are 4 possible combinations, arising out of (i) and (ii) in terms of (plus/minus, plus/minus). We use two of the combinations and derive
(a) $(y_1 = 5.32, y_2 = 2.56, y_3 = 2.62)$; (b) $(y_1 = 4.45, y_2 = 1.67, y_3 = 4.38)$.

Remark 4.2.1 This result can be extended naturally for any general sample size n, by referring to

(a) Helmert's Orthogonal Transformation Matrix of order $n \times n$ and
(b) Polar Transformation—generally used to deduce the sampling distribution of Chi-square.

Note that $\text{SS}(X)/\sigma^2$ follows Chi-square distribution with $(n-1)$ df.

It is educational and instructional to understand the importance of suitable transformations such as
(1) $|Y_1 - Y_2|/\sqrt{2} = \sqrt{\text{SS}(X)}\cos\theta_1$,
(2) $|Y_1 + Y_2 - 2Y_3|/\sqrt{6} = \sqrt{\text{SS}(X)}\sin\theta_1\cos\theta_2$,
(3) $|Y_1 + Y_2 + Y_3 - 3Y_4|/\sqrt{12} = \sqrt{\text{SS}(X)}\sin\theta_1\sin\theta_2\cos\theta_3, \ldots$,
$((n-2))\ |Y_1 + Y_2 + Y_3 + \cdots + Y_{(n-2)} - (n-2)Y_{(n-1)}|/\sqrt{(n-1)(n-2)}$
$= \sqrt{\text{SS}(X)}\sin\theta_1\sin\theta_2\ldots\cos\theta_{(n-2)}$,
$((n-1))\ |Y_1 + Y_2 + Y_3 + \cdots + Y_{(n-1)} - (n-1)Y_{(n)}|/\sqrt{n(n-1)}$
$= \sqrt{\text{SS}(X)}\sin\theta_1\sin\theta_2\ldots\sin\theta_{(n-2)}$.

This is obtained by a combination of Helmert's Orthogonal Transformation and Polar Transformation. Here θ's are independent, each ranging in $[0, \pi/2]$. In reality, we select the θ's independently and compute the right-hand side expression values [in terms of observations y's], using the observed value $\text{ss}(x)$ of $\text{SS}(X)$. Then the left-hand side expressions are known. For each one, we consider both the options as 'plus' or 'minus'. Then at the last leg, we use the known value of $\bar{x}$ and the property of Helmert's Orthogonal Matrix to solve for the y's. These y's are the realizations of Y's which are then distributionally equivalent to the X's.

Example 4.2.3 Consider $n = 6$ and suppose $\bar{x} = 2.40$ and $\text{ss}(x) = 36$. We will generate all the 6 observations, viz. $y_1, y_2, \ldots, y_6$ which we claim to be the realization of Y's and these Y's are distributionally equivalent to the X's. Since $n = 6$, we will make random choices of $\theta_1, \theta_2, \theta_3, \theta_4$—each in the range $[0, \pi/2]$ and then compute their sine and cosine function values. Here θ's are expressed in degrees: $\theta_1 = 29, \theta_2 = 79, \theta_3 = 56, \theta_4 = 11$.

(i) $\text{sine}\,29 = 0.4848$, $\text{cosine}\,29 = 0.8746$,
(ii) $\text{sine}\,79 = 0.9816$, $\text{cosine}\,79 = 0.1908$,
(iii) $\text{sine}\,56 = 0.8290$, $\text{cosine}\,56 = 0.5592$,
(iv) $\text{sine}\,11 = 0.1908$, $\text{cosine}\,11 = 0.9816$.

(1) $|y_1 - y_2|/\sqrt{2} = 60.8746 = 5.2476$,
(2) $|y_1 + y_2 - 2y_3|/\sqrt{6} = 6 \times 0.4848 \times 0.1908 = 0.5550$,
(3) $|y_1 + y_2 + y_3 - 3y_4|/\sqrt{12} = 6 \times 0.4848 \times 0.9816 \times 0.5592 = 1.5967$,
(4) $|y_1 + y_2 + y_3 + y_4 - 4y_5|/\sqrt{20} = 6 \times 0.4848 \times 0.9816 \times 0.8290 \times 0.9816 = 2.3235$,
(5) $|y_1 + y_2 + y_3 + y_4 + y_5 - 5y_6|/\sqrt{30} = 6 \times 0.4848 \times 0.9816 \times 0.8290 \times 0.1908 = 0.4516$.
On the top of these equations, we have the one based on the sample mean. This yields
(6) $|y_1 + y_2 + y_3 + y_4 + y_5 + y_6|/\sqrt{6} = [x_1 + x_2 + \cdots + x_6]/\sqrt{6} = (62.40)/2.4495 = 5.8787$.

From the above, we derive, using all 'plus' values,

$$y_1 = 7.4,\ y_2 = 0.02,\ y_3 = 3.01,\ y_4 = 1.62,\ y_5 = 0.40,\ y_6 = 1.99.$$

It is readily verified that (i) $\bar{y} = 2.40 = \bar{x}$ and (ii) $\text{ss}(y) = 36.00 = \text{ss}(x)$.

We realize that there are altogether $2^5 = 32$ solutions available for the y's by choosing the signs (plus/minus) of the linear functions of the y's in the above equations (1) to (5). We display below a few other solutions:

(1) Case of plus/minus/minus/plus/minus/plus:

$$y_1 = 5.86,\ y_2 = 1.56,\ y_3 = 2.83,\ y_4 = 4.22,\ y_5 = 0.24,\ y_6 = 2.81;$$

(2) Case of plus/plus/minus/minus/plus/plus:

$$y_1 = 5.44,\ y_2 = 1.98,\ y_3 = 1.05,\ y_4 = 3.35,\ y_5 = 4.56,\ y_6 = 1.99;$$

(3) Case of minus/plus/minus/plus/minus/plus:

$$y_1 = 1.11,\ y_2 = 6.31,\ y_3 = 1.92,\ y_4 = 4.22,\ y_5 = 0.24,\ y_6 = 2.81.$$

In the next section, we discuss about linear regression models with known/unknown homoscedastic error variance σ^2 and examine the nature of sufficient experiments.

4.3 The Notion of Sufficient Experiments in Linear Regression Set-Up

We start with the following result in the context of linear regression with known homoscedastic variance σ^2.

Theorem 4.3.1 *If $Y_{+1} = Y(+1)$ and $Y_{-1} = Y(-1)$ are independent univariate normal variables with parameters $(\alpha + \beta, \sigma^2)$ and $(\alpha - \beta, \sigma^2)$, then $Y(+1)$ and $Y(-1)$ are jointly sufficient statistics for generating any Y_x whenever $-1 \leq x \leq 1$ and σ^2 is known. Further to this, one can generate any pair $Y_{(X_1)}$ and $Y_{(X_2)}$ based on $X(+1)$ and $X(-1)$ and these two will be uncorrelated if and only if $-1 \leq X_1 < X_2 \leq 1$ and $X_1 + X_2 = 0$.*

Proof To generate Y_x, we use $\hat{\alpha}$ and $\hat{\beta}$ based on $Y(+1)$ and $Y(-1)$ and start with $\hat{\alpha} + \hat{\beta}x$ which is unbiased for $\alpha + \beta x$. For this to hold, we do not need any restriction on x. Next note that, explicitly written,

$$\hat{\alpha} + \hat{\beta}x = Y(+1)[1+x]/2 + Y(-1)[1-x]/2.$$

Accordingly,

$$\begin{aligned}\text{Var}\,[Y(+1)[1+x]/2 + Y(-1)[1-x]/2] &= (1+x^2)]\sigma^2/2 < \sigma^2\\ \text{whenever}\quad -1 < x < 1.\end{aligned}$$

In order that we can generate Y_x for any $-1 < x < 1$, based on $Y(+1)$ and $Y(-1)$, we need to make up for the deficit in error variance σ^2 by the amount $(1-x^2)\sigma^2/2$. This is possible since the error variance σ^2 is assumed to be known and $1 < x < 1$. We generate an observation from $N(0, (1-x^2)\sigma^2/2)$, say $Z(x)$. Then $Y(+1)[1+x]/2 + Y(-1)[1-x]/2 + Z(x)$ serves our purpose. It provides an observation which is distributionally equivalent to $Y(x)$. Hence the first part of the claim is established. We must note in passing that the condition $-1 < x < 1$ is essential for this result to hold.

Next, we take up two distinct points $-1 < x_1 < x_2 < 1$ and generate the corresponding responses

$$Y(+1)[1+x_1]/2 + Y(-1)[1-x_1]/2 + Z(x_1),$$

and

$$Y(+1)[1+x_2]/2 + Y(-1)[1-x_2]/2 + Z(x_2),$$

where $Z(x_i)$s are independently distributed as $N(0, (1-x_i^2)\sigma^2/2), i = 1, 2$.

Now we compute an expression for the covariance between the two generators and equate the same to zero. This suggests

$$(1+x_1)(1+x_2) + (1-x_1)(1-x_2) = 0 = 1 + x_1x_2.$$

This implies $x_1 = -1, x_2 = +1$ which is not strictly valid, since we assumed $-1 < x_1 < x_2 < 1$. However, we can make an alternative choice of the errors $Z(x_i)$'s. We may set $x_1 + x_2 = 0$ and choose $Z(x_2) = -Z(x_1)$ in the above. This is possible since the error variance for generating the Zs involves x_i's through x_i^2s only. Then the covariance term would simplify to

$$(1 + x_1)(1 + x_2) + (1 - x_1)(1 - x_2) - 2(1 - x_1^2) = 0.$$

Hence we can achieve our goal of generating two independent responses under the given linear regression model at the points (x_1, x_2) when $x_1 + x_2 = 0$.

It is clear that independent choices of Z_1 and Z_2 will not lead to any positive result. Further, if we choose $Z_1 = Z_2$, then the covariance term would simplify to $x_1x_2 + x_1^2 = 0$ which simplifies to $x_1 = 0$ or $x_1 + x_2 = 0$. Barring the case of $x_1 = 0$, we find that $x_1 + x_2 = 0$ is also a necessary condition.

Hence the claim is established.

Remark 4.3.1 Since $Y(+1)$ and $Y(-1)$ jointly capture the entire information for α and β, if there are more than 2 observations, we will circulate between $Y(+1)$ and $Y(-1)$ alternately. Thus for 3 observations, we can use $Y(+1)^{(1)}, Y(+1)^{(2)}, Y(-1)$. Then we can generate simultaneously three Y-observations located at three distinct X-values $(-1 \leq X_1 < X_2 < X_3 \leq 1)$ and these generated Y-values will be uncorrelated if and only if $X_1 + X_2 + X_3 = 1$. This result is stated and proved below.

Theorem 4.3.2 *If $Y(+1)^{(1)}$ and $Y(+1)^{(2)}$ are two independent observations—both generated at $X = 1$ and further, $Y(-1)$ is a third independent observation generated at $x = -1$, then [based on these three generated observations] we can generate three observations at three distinct points $-1 < x_1 < x_2 < x_3 < 1$—having the same law of distribution. Moreover, these three newly generated observations will be uncorrelated if and only if $x_1 + x_2 + x_3 = 1$.*

Proof We set

$$U = [Y(+1)^{(1)} + Y(+1)^{(2)}]/2, V = Y(-1). \tag{4.3.1}$$

Then

$$\hat{\alpha} = (U + V)/2, \hat{\beta} = (U - V)/2. \tag{4.3.2}$$

Therefore, generating an error function W from $N(0, \sigma^2)$ and using $Q(x) = (5 + 2x - 3x^2)/8 = (5 - 3x)(1 + x)/8 > 0$, we may assert that

$$(U + V)/2 + x(U - V)/2 + W\sqrt{Q(x)} = U(1 + x)/2 + V(1 - x)/2 + W\sqrt{Q(x)} \tag{4.3.3}$$

serves as an unbiased predictor for Y_x for any $-1 < x < 1$. Incidentally, our choice of $Q(x)$ reveals that variance of the expression in (4.3.3) is given by

$$(1 + x)^2/8 + (1 - x)^2/4 + Q(x) = 1, \text{ as desired.}$$

Again, for three distinct values of x, say $-1 < x_1 < x_2 < x_3 < 1$, we can generate the corresponding unbiased predictors as

$$U(1 + x_i)/2 + V(1 - x_i)/2 + W_i\sqrt{Q(x_i)};\ i = 1, 2, 3. \qquad (4.3.4)$$

Note that in (4.3.4), the W_i's are each marginally $N(0, \sigma^2), i = 1, 2, 3$. We will now specify their pairwise covariances such that the three predictors follow trivariate normal distribution and are pairwise uncorrelated. We write $W_i^* = W_i\sqrt{Q(x_i)}, i = 1, 2, 3$. We comfortably drop σ^2 from the expressions, as and when suitable/ convenient.

Note that pairwise zero covariance will be realized when

$$(1 + x_i)(1 + x_j)/8 + (1 - x_i)(1 - x_j)/4 + \text{Cov}(W_i, W_j) = 0, i \neq j = 1, 2, 3.$$

This suggests

$$\text{Cov}\ (W_i, W_j) = (1/8)[(-3 + x_i + x_j - 3x_i x_j)].$$

On the other hand,

$$V(W_i) = Q(x_i) = (5 + 2x_i - 3x_i^2)/8.$$

Therefore, required form of the dispersion matrix of the predictors is given by [dropping the factor (1/8) and σ^2 conveniently],

$$\begin{aligned}\Sigma = [&(5 + 2x_1 - 3x_1^2)\ \ (-3 + x_1 + x_2 - 3x_1x_2)\ \ (-3 + x_1 + x_3 - 3x_1x_3);\\ &(-3 + x_1 + x_2 - 3x_1x_2)\ \ (5 + 2x_2 - 3x_2^2)\ \ (-3 + x_2 + x_3 - 3x_2x_3);\\ &(-3 + x_1 + x_3 - 3x_1x_3)\ \ (-3 + x_2 + x_3 - 3x_2x_3)\ \ (5 + 2x_3 - 3x_3^2)],\end{aligned}$$

which can be expressed as

$$\Sigma = 8\mathbf{I} - \mathbf{A};\ \mathbf{A} = 3\mathbf{J} + 3\mathbf{x}\mathbf{x}' - \left[\mathbf{x}\mathbf{1}' + \mathbf{1}\mathbf{x}'\right].$$

Now note that when $x_1 + x_2 + x_3 = 1$, $\Sigma\mathbf{1} = \mathbf{0}$ so that $\mathbf{A}$ has one eigenvalue equal to 8. Furthermore, $\mathbf{A}$ can be expressed as

$$\begin{aligned}\mathbf{A} = [&(1 + x_1\ \ 1 + x_2\ \ 1 + x_3)\ \ (1 + x_1\ \ 1 + x_2\ \ 1 + x_3)\\ &2(1 - x_1\ \ 1 - x_2\ \ 1 - x_3)\ \ (1 - x_1\ \ 1 - x_2\ \ 1 - x_3)]\end{aligned}$$

which suggests that its rank is 2. Also, when $x_1 + x_2 + x_3 = 1$,

$$\text{Trace}\ (\mathbf{A}) = 7 + 3\sum x_i^2 = 8 + \left[3\sum x_i^2 - 1\right].$$

Therefore, the other nonzero eigenvalue of **A** is $(3\sum x_i^2 - 1) > 0$. Finally, we derive

$$\left[0, 8, 8 - \left(3\sum x_i^2 - 1\right)\right] = \left[0, 8, 3\left(3 - \sum x_i^2\right) > 0\right]$$

as the eigenvalues of Σ. Hence the matrix is non-negative definite, and we are in a position to generate three uncorrelated random variables as prototypes of the Y's at $x = x_1, x_2, x_3;\ 1 < x_1 < x_2 < x_3 < 1$, when $x_1 + x_2 + x_3 = 1$.

Remark 4.3.2 Alternately, for 3 observations, we can also use $Y(-1)^{(1)}$, $Y(-1)^{(2)}$, $Y(+1)$. Then the statement made in Theorem 3.2 will be changed accordingly and we need the condition $x_1 + x_2 + x_3 = -1$ for the validity of the claim.

Remark 4.3.3 In case σ^2 is unknown in a linear regression model, we need three observations to accommodate the three parameters. Here again sufficiency holds in terms of the extreme observations.

Theorem 4.3.3 *If $Y(+1)^{(1)}$ and $Y(+1)^{(2)}$ are two independent observations—both generated at $X = +1$ and further, $Y(-1)$ is a third independent observation generated at $X = -1$, then we can unbiasedly predict Y_x for any $-1 < x < 1$ even when the error variance σ^2 is unknown. Further to this, we can unbiasedly predict Y_{x_1} and Y_{x_2} for suitably chosen $-1 < x_1 < x_2 < 1$ and these predictors will be uncorrelated for many choices of the pair $-1 < x_1 < x_2 < 1$.*

Proof Set $Y_x = \alpha + \beta x + e_x$ and, further,

$$U = \left[Y(+1)^{(1)} + Y(+1)^{(2)}\right]/2,\ V = Y(-1),\ W = \left[Y(+1)^{(1)} - Y(+1)^{(2)}\right]/\sqrt{2}. \tag{4.3.5}$$

Note that W is distributed as $N(0, \sigma^2)$. Further,

$$\hat{\alpha} = (U + V)/2, \hat{\beta} = (U - V)/2; \hat{\sigma^2} = W^2. \tag{4.3.6}$$

Therefore, for $-1 < x < 1$, using $Q(x) = [5 - 2x - 3x^2]/8 > 0$, we may assert that

$$(U + V)/2 + x(U - V)/2 + W\sqrt{Q(x)} \tag{4.3.7}$$

serves as an unbiased predictor for Y_x for any $-1 < x < 1$. Again, for two distinct values of x, say $-1 < x_1 < x_2 < 1$, we can generate the corresponding y-values as

$$(U + V)/2 + x_1(U - V)/2 + W\sqrt{Q(x_1)};\ (U + V)/2 + x_2(U - V)/2 - W\sqrt{Q(x_2)}. \tag{4.3.8}$$

Note that W has mean zero. We now evaluate an expression for the covariance between the two generators. Routine computations yield, upon equating to zero,

$$(1 + x_1)(1 + x_2)/8 + (1 - x_1)(1 - x_2)/4 = Q(x_1)Q(x_2). \quad (4.3.9)$$

We now solve for (x_1, x_2) by referring to $-1 < x_1 < x_2 < 1$. We find interesting geometric interpretation of the solution space—apart from a large number of solutions. It transpires that $-0.365 < x_1 \neq x_2 < 0.298$. Some special pairs of values are listed here.

$$\begin{aligned}&(0, 1/5); (0, -1/3); (0.1, 0.1158); (0.1, -0.39)(0.2, -0.006), (0.2, -0.42);\\ &(0.25, -0.97), (0.25, -0.415);\\ &(-0.1, -0.258), (-0.1, 0.262); (-0.2, -0.161), (-0.2, 0.302);\\ &(-0.3, -0.0223), (-0.3, 0.3062). \quad (4.3.10)\end{aligned}$$

Next we take up similar study for a quadratic regression model with/without the intercept term so that still suitably chosen points serve the purpose in terms of sufficiency. We again assume σ^2 to be known.

Theorem 4.3.4 *We consider a quadratic regression model [with the intercept term] of the form $Y_x = \alpha + \beta x + \gamma x^2 + e_x$, for $-1 \leq x \leq 1$ and with known homoscedastic error variance σ^2. Then the observations at three points $Y(+1)$, $Y(-1)$ and $Y(0)$ are jointly sufficient for generating an observation at any point x, $-1 < x < 1$. However, for generating a pair of observations corresponding to two distinct x-values, say x_1 and x_2, having zero-correlation, there is no interesting general result. We deduced that*

(i) when $x_1 + x_2 = 0$, this is possible if and only if x_1, assumed to be positive, $\geq 1/\sqrt(3)$;
(ii) when $x_1^2 + x_2^2 = 1$, this is possible if and only if x_1, assumed to be positive, satisfies the inequality: $0.1691 < x_1 < 0.9856$.

Proof (First Part) : We note that

$$\hat{\alpha} = y(0); \hat{\beta} = [y(+1) - y(-1)]/2; \hat{\gamma} = [y(+1) + y(-1) - 2y(0)]/2.$$

Then, for any other value of x, $-1 < x \neq 0 < 1$, the predicted value for Y_x can be derived by starting with the computation of $V[\hat{\alpha} + \hat{\beta}x + \hat{\gamma}x^2]$ which is the same as computation of

$$\begin{aligned}&V\ \left[y(0) + x[y(+1) - y(-1)]/2 + x^2[y(+1) + y(-1) - 2y(0)]/2\right]\\ &= V\ [y(0)(1 - x^2) + y(+1)x(1 + x)/2 - y(-1)x(1 - x)/2]\\ &= \left[(1 - x^2)^2 + x^2(1 + x)^2/4 + x^2(1 - x)^2/4\right] \times \sigma^2\\ &= \left[(1 - x^2)^2 + x^2(1 + x^2)/2\right] \times \sigma^2\\ &= \left[1 - 3x^2(1 - x^2)/2\right] \times \sigma^2 = \sigma^2\phi(x), \text{ say } \leq \sigma^2.\end{aligned}$$

Note that $5/8 \leq \phi(x) \leq 1$. So we can blend an independent error function which is normally distributed with variance $(1 - \phi(x)) \times \sigma^2$. That will provide an unbiased predictor for Y_x with variance exactly equal to σ^2. Note that this holds for any $-1 < x < 1$.

Proof (Second Part) : Note that for each $x_i, i = 1, 2$, we will blend $\hat{\alpha} + \hat{\beta}x_i + \hat{\gamma}x_i^2$ with an independent error component W_i which is normally distributed with variance $(1 - \phi(x_i))\sigma^2$. Since $\hat{\alpha}$, $\hat{\beta}$, and $\hat{\gamma}$ are common to both the predictors, an expression for the covariance between the two predictors is given by

$$(1 - x_1^2)(1 - x_2^2) + x_1x_2\left[(1 + x_1)(1 + x_2) + (1 - x_1)(1 - x_2)\right]/4 + \mathrm{Cov}(W_1, W_2).$$

We have two options now.

(i) We may set W_1, W_2 to be themselves uncorrelated. That means, we may search for (x_1, x_2) combinations for which the above expression involving (x_1, x_2) is zero.

It is found that plenty of choices of (x_1, x_2) are available for this to materialize. We list several such choices below.

$$(-0.9, 0.5867), (-0.8, 0.8309), (-0.7, 0.8760), (-0.6, 0.8970), (-0.5, 0.9135),$$

$$(-0.4, 0.9279), (-0.3, 0.9426), (-0.2, 0.9588), (-0.1, 0.9775), (0, 1.00).$$

Otherwise,

(ii) We may attach a covariance term to the joint distribution of the error terms. That means, we set

$$\mathrm{Cov}\,(W_1, W_2) = -\left[(1 - x_1^2)(1 - x_2^2) + x_1x_2\left[(1 + x_1)(1 + x_2) + (1 - x_1)(1 - x_2)\right]/4\right]$$

which simplifies to

$$-1 + x_1^2 + x_2^2 - x_1x_2/2 - 3x_{21}x_2^2/2 = -\left[1 - (x_1^2 + x_2^2) + x_1x_2(1 + 3x_1x_2)/2\right].$$

However, this will work if and only if the associated dispersion matrix of (W_1, W_2), so formed, is an nnd matrix. For this we need to assert that for a given choice of (x_1, x_2), the underlying dispersion matrix

$$[1 - \phi(x_1), -[1 - (x_1^2 + x_2^2) + x_1x_2(1 + 3x_1x_2)/2]; -[1 - (x_1^2 + x_2^2) + x_1x_2(1 + 3x_1x_2)/2], 1 - \phi(x_2)]$$

is nnd. The diagonals are already non-negative. So we need non-negativity condition on the 2×2 determinant. This suggests

$$9 \times x_1^2 x_2^2 (1 - x_1^2)(1 - x_2^2)/4 \geq \left[1 - (x_1^2 + x_2^2) + x_1 x_2 (1 + 3x_1 x_2)/2\right]^2.$$

At this stage, we may specialize to two different cases.
Case I : $-1 < x_2 < 0 < x_1 < 1; x_1 + x_2 = 0$. We set $x_1 = t, x_2 = -t, 0 < t < 1$. In this case, the above expression simplifies to $(-1)(1 - t^2)^2(1 - 3t^2) \geq 0$. This is equivalent to $t \geq 1/\sqrt{(3)}$.
Case II : $x_1^2 + x_2^2 = 1, -1 < x_2 < 0 < x_1 < 1$. We set $x_1^2 = t, 0 < t < 1$. Then non-negativity amounts to the condition: $t(1 - t) > 1/36$, i.e., $0.0286 < t < 0.9714$. This amounts to: $0.1691 < x_1 < 0.9856$.

The special cases thus lead to some existence results for the pair (x_1, x_2) in order to generate two uncorrelated predictors.

We may now supplement the above results by considering the case of unknown error variance σ^2.

Theorem 4.3.5 *Consider a quadratic regression model [with the intercept term] of the form $Y_x = \alpha + \beta x + \gamma x^2 + e_x$, for $-1 \leq x \leq 1$ and with unknown but homoscedastic error variance σ^2. Then a total of four observations at three points $Y(+1), Y(-1)$ and $(Y(1)(0), Y(2)(0))$ are jointly sufficient for generating an observation at any point $x, -1 < x < 1$. In other words, in case the error variance σ^2 is unknown, we may repeat $Y(0)$ twice, i.e., independently generate $Y(1)(0)$ and $Y(2)(0)$. Then their average will produce $\hat{\alpha}$, and their difference will produce an error function. This error function can now be suitably used to provide an unbiased predictor for $Y(x)$ for any $-1 < x < 1$.*

We skip proof of this result.

Theorem 4.3.6 *Consider a quadratic regression model without the intercept term of the form $Y_x = \beta x + \gamma x^2 + e_x$, for $-1 \leq x \leq 1$ and with known homoscedastic error variance σ^2. Then again $Y(+1)$ and $Y(-1)$ are jointly sufficient. Further, in order to generate a pair of uncorrelated observations at x_1 and x_2, we need to separate out two cases:*

(i) $x_1 + x_2 = 0$ which forces the trivial choice $x_1 = -1, x_2 = +1$.
(ii) $x_1 + x_2 \neq 0$. Here we have a host of positive results and we may describe them by referring to the ratio $p = x_1/x_2$, assumed to be positive.

Proof Under the stated model,

$$\hat{\beta} = [Y(+)Y(-)]/2; \hat{\gamma} = [Y(+1) + Y(-1)]/2.$$

In effect, therefore, we can produce unbiased predictor for Y_x provided $V[\hat{\beta}x + \hat{\gamma}x^2] \leq \sigma^2$. In this case, the variance expression turns out to be $(x^2 + x^4)\sigma^2/2$ which is less than or equal to σ^2 whenever $-1 < x < 1$. The difference $(2 - x^2 - x^4)\sigma^2/2$

can be attributed to the error variance by bringing in an independent error component W as usual.

Again, for a pair of x-values, say $-1 < x_1 < x_2 < 1$, we may associate two separate error components W_1 and W_2 as usual. Next we compute covariance between the two predictors as

$$[x_1x_2(1 + x_1)(1 + x_2) + x_1x_2(1 - x_1)(1 - x_2)]\,\sigma^2/4 = [x_1x_2(1 + x_1x_2)]\,\sigma^2/2.$$

Without any correlation between the two error components W_1 and W_2, the above expression would be 0 if and only if $x_1x_2 = 1$ i.e., $x_1 = 1, x_2 = 1$ which is not the case here. Therefore, we will introduce a covariance term between W_1 and W_2 and it is evident that $\mathrm{Cov}(W_1, W_2) = (-1)[x_1x_2(1 + x_1x_2)]\sigma^2$.

This suggests a form of the dispersion matrix for joint distribution of W_1 and W_2 as [dropping σ^2)

$$\left[2 - x_1^2(1 + x_1^2), -x_1x_2(1 + x_1x_2); -x_1x_2(1 + x_1x_2), 2 - x_2^2(1 + x_2^2)\right].$$

We now impose the condition of non-negativity of this matrix for suitably chosen pair of (x_1, x_2)-values. Diagonal elements are non-negative anyway. The determinant is non-negative if and only if

$$4 - 2(x_1^2 + x_2^2) - 2(x_1^4 + x_2^4) + x_1^2x_2^2(x_1 - x_2)^2 \geq 0.$$

For some choices of $(x_1, x_2), -1 < x_1 < x_2 < 1$, it transpires that the determinantal value is negative. On the other hand, there are plenty of choices of (x_1, x_2) for which the above non-negativity condition holds. Moreover, if for a choice of both positive values, i.e., $0 < x_1 < x_2 < 1$, non-negativity holds, then it holds for all the four combinations $(\pm x_1, \pm x_2)$. For example, we have $(\pm 0.9, \pm 0.45)$ as affirmative choices.

Finally, we take up a similar study when σ^2 is unknown. Naturally, $Y(0)$ is an error function and it can be used straightaway along with $Y(+1)$ and $Y(-1)$. This time we can take care of the difference amount $(2 - x^2 - x^4)\sigma^2/2$ by referring to $\phi(x)Y(0)$ where $\phi^2(x) = (2 - x^2 - x^4)/2$. For generating two uncorrelated predictors corresponding to (x_1, x_2), we may use the same error $Y(0)$ with opposite signs:

$$\hat{\beta}x_1 + \hat{\gamma}x_1^2 + \phi(x_1)Y(0), \hat{\beta}x_1 + \hat{\gamma}x_1^2 - \phi(x_2)Y(0).$$

We need to work out an expression for the covariance between the two generators and equate the same to zero. This yields $[x_1x_2(1 + x_1x_2)] = \phi(x_1)\phi(x_2)$ and hence to the equation

$$4x_1^2x_2^2(1 + x_1x_2)^2 = (2 - x_1^2 - x_1^4)(2 - x_2^2 - x_2^4)$$

which simplifies, in the particular hypothetical case of $x_1 = x_2 = t$ say, to $3t^4(1 + t^2)^2 + 4t^2(1 + t^2)^2 = 4$. This has a nice solution to t^2 as $t^2 = 0.4574$ implying thereby that $t = 0.6763$. Because of continuity, we can derive acceptable solutions for the pair of x-values. For example, $x_1 = 0.69$, $x_2 = 0.86$ is a solution [in an approximate sense] for which we can provide unbiased and uncorrelated predictors of the underlying responses.

4.4 The Notion of Sufficient Experiments in Bivariate and Trivariate Normal Populations

Theorem 4.4.1 *Consider a bivariate normal distribution model for (X_1, X_2) with known means and known variances but with an unknown correlation ρ. Let there be a competing distribution involving (Y_1, Y_2) with means and variances all known but having the correlation coefficient $c\rho$ where $c > 1$ is a known constant. Then (Y_1, Y_2) is sufficient for (X_1, X_2) in the following sense:*

Given (Y_1, Y_2), we can generate another pair (Y_1^*, Y_2^*) having the same distribution as (X_1, X_2).

Generalizing this result, we may deduce

Corollary 4.4.1 *Consider a bivariate normal distribution model for (X_1, X_2) with known means and known variances but with an unknown correlation $c_1\rho$. Let there be a competing distribution involving (Y_1, Y_2) with means and variances all known but having the correlation coefficient $c_2\rho$. If both (c_1, c_2) are known and $c_1 < c_2$, then (Y_1, Y_2) is sufficient for (X_1, X_2) in the following sense:*

Given (Y_1, Y_2), we can generate another pair (Y_1^*, Y_2^*) having the same distribution as (X_1, X_2).

In terms of information content on ρ in a bivariate normal population with known means and variances, it turns out that a comparison of Fisher Information can be readily made in some cases. Note that $I(\rho) = \frac{(1+\rho^2)}{(1-\rho^2)^2}$ is the expression for Fisher Information on ρ in a bivariate normal population with known means and variances.

Theorem 4.4.2 *Consider a bivariate normal distribution model for (X_1, X_2) with known means and known variances but with an unknown correlation $c_1\rho$. Let there be a competing distribution involving (Y_1, Y_2) with means and variances all known but having unknown correlation coefficient $c_2\rho$. If both (c_1, c_2) are known and $c_1 < c_2$, then (Y_1, Y_2) is more informative than (X_1, X_2) in the sense of more information content as measured by Fisher's Information.*

Remark 4.4.1 There is another general result involving two bivariate normal populations and the criterion of Fisher Information. It is stated like this:

Consider a bivariate normal distribution model for (X_1, X_2) with known means and known variances but with an unknown correlation ρ. Let there be a competing distribution involving (Y_1, Y_2) with means and variances all known but having unknown correlation coefficient ρ^s, $s > 0$. Then (Y_1, Y_2) is more informative than

(X_1, X_2) in the sense of more information content as measured by Fisher's Information whenever $0 < s < 1$.

Regarding Trivariate Normal Populations, we have similar results on Sufficiency.

Theorem 4.4.3 *Consider a trivariate normal distribution model for* (X_1, X_2, X_3) *with known means and known variances and with the same but unknown pairwise correlation coefficient* ρ. *Let there be a competing distribution involving* (Y_1, Y_2, Y_3) *with means and variances all known and again having equal but unknown pairwise correlation coefficient* $c\rho$, $c > 0$. *Then* (Y_1, Y_2, Y_3) *is sufficient for* (X_1, X_2, X_3) *whenever* $c > 1$.

From the above, we may deduce the following.

Corollary 4.4.2 *Consider a trivariate normal distribution model for* (X_1, X_2, X_3) *with known means and known variances and with the same but unknown pairwise correlation coefficient* $c_1\rho$. *Let there be a competing distribution involving* (Y_1, Y_2, Y_3) *with means and variances all known and again having equal but unknown pairwise correlation coefficient* $c_2\rho$. *Then* (Y_1, Y_2, Y_3) *is sufficient for* (X_1, X_2, X_3) *whenever* $c_2 > c_1$.

Additional results on the comparison of trivariate normal populations are stated below.

Theorem 4.4.4 *Consider a trivariate normal distribution model for* (X_1, X_2, X_3) *with known means and known variances and with the unknown pairwise correlation coefficients* $\rho_{12}, \rho_{13}, \rho_{23}$. *Let there be a competing distribution involving* (Y_1, Y_2, Y_3) *with means and variances all known and having unknown pairwise correlation coefficients* $c\rho_{12}, c\rho_{13}, c\rho_{23}$. *Then* (Y_1, Y_2, Y_3) *is sufficient for* (X_1, X_2, X_3) *whenever* $c > 1$.

Theorem 4.4.5 *Consider a trivariate normal distribution model for* (X_1, X_2, X_3) *with known means and known variances and with the unknown pairwise correlation coefficients* $\rho_{12}, \rho_{13}, \rho_{23}$. *Let there be a competing distribution involving* (Y_1, Y_2, Y_3) *with means and variances all known and having unknown pairwise correlation coefficients* $c_1\rho_{12}, c_2\rho_{13}, c_3\rho_{23}$. *Then* (Y_1, Y_2, Y_3) *is sufficient for* (X_1, X_2, X_3) *whenever* $c_1 < c_2c_3, c_2 < c_1c_3, c_3 < c_1c_2$.

This is a non-trivial result, and we propose to provide a proof of this claim. Without any loss of generality, we assume each of the X_i's and Y_i's to have mean zero and variance unity.

For positive real numbers a_1, a_2, a_3, we make a transformation from Y_i's to Y_i^*'s by introducing three iid $N(0, 1)$ variables $U_i, i = 1, 2, 3$ and by defining

$$Y_i^* = \frac{\sqrt{(a_i)}Y_i + U_i}{\sqrt{(a_i)} + 1}, i = 1, 2, 3.$$

Clearly, each Y_i^* is $N(0, 1)$ and, moreover, pairwise covariances are

$$\text{Cov}(Y_1^*, Y_2^*) = \frac{\sqrt{(a_1a_2)}c_1\rho_1}{\sqrt{(a_1+1)(a_2+1)}}$$

and the like.

We will be equating the covariances to those of the pairs of X's, i.e., to ρ_1, ρ_2, ρ_3. In effect, we obtain the three equations as $\frac{\sqrt{a_1a_2}c_1}{\sqrt{(a_1+1)(a_2+1)}} = 1$ and the like. Upon simplification, we derive

$$a_1 = \frac{c_3}{c_1c_2 - c_3}, a_2 = \frac{c_2}{c_1c_3 - c_2}, a_3 = \frac{c_1}{c_2c_3 - c_1}.$$

The conditions stated above provide the solution for every choice of (c_1, c_2, c_3), irrespective of the choice of the ρ_i's.

4.5 The Notion of Sufficient Experiments in One Way ANOVA Models with Random Effects

The set-up is described as follows. There are v treatments with random effects with mean zero and constant known variance $\sigma_t^2 > 0$. There are all together n experimental plots and treatments are randomly allocated following the replication numbers $(n_1, n_2, \ldots, n_v)$. The overall mean θ is assumed to be unknown, while treatment variance (σ_t^2) and error variance (σ_e^2) are both assumed to be known. The problem is to compare different allocation designs and, if possible, to arrive at the optimal allocation design in the sense of sufficient experiment. This problem has a long history and it traces back to 1970s.

It will be established that of all the allocations for given n, total number of experimental plots and for arbitrary v, the number of treatments (with random effects), $1 \leq v \leq n$, the one defined as $A(n, n; 1, 1, \ldots, 1)$ is sufficient for all others. In a series of Lemmas, we will arrive at this result. In a way, under a random effects model with unknown overall model mean θ and known variance components, the best bet is to use n random treatments—each one with a single replication. These can then be suitably combined to provide data arising out of any other allocation.

Lemma 4.5.1 *In the set-up of one-way random effects model with unknown mean θ and known variance components $[\sigma_t^2$ and $\sigma_e^2]$, for a set of at most k randomly chosen treatments and k randomly chosen plots, the allocation $A(k, k; 1, 1, \ldots, 1)$ is sufficient for the allocation $A(k, 1; k)$.*

Proof From the allocation $A(k, k; 1, 1, \ldots, 1)$, the observations undergo the linear model

$$y_i = \theta + t_i + e_i; i = 1, 2, \ldots, k,$$

where t_i's are iid random treatment effects with means zero's and each with the same known variance σ_t^2. Moreover, the errors e_i's are iid with mean zeros and known variance σ_e^2. Clearly,

$$\hat{\theta} = \bar{y}, \text{Var}(\bar{y}) = \left[\sigma_t^2 + \sigma_e^2\right]/k.$$

On the other hand, the observations underlying the allocation $A(k, 1; k)$ follow the linear model

$$u_i = \theta + t_1 + e_i, i = 1, 2, \ldots, k$$

so that the u_i's have each mean θ, variance $(\sigma_t^2 + \sigma_e^2)$ and pairwise covariance σ_t^2. Our purpose is to generate an equivalent version of u_i's from the y_i's.

We first generate, independent of the y's,

(i) univariate normal random variable z distributed as $N(0, (1 - 1/k)\sigma_t^2)$;
(ii) k-variate singular normal random vector $\mathbf{S}$ distributed as $N_k(0, (I - J/k)\sigma_e^2)$.
Based on $\bar{y}$, z and $\mathbf{S}$, we now generate a set of k random observations defined as $v_i = \bar{y} + z + S_i, i = 1, 2, \ldots, k$.

Since the components of v_i's are normally distributed, it follows that v_i's have a joint k-variate normal distribution, each with mean θ, variance $\sigma_t^2 + \sigma_e^2$ and with pairwise covariance σ_e^2.

In other words, v_i's reproduce an equivalent version of the u_i's underlying the allocation $A(k, 1; k)$.

This establishes Lemma 4.5.1.

Theorem 4.5.1 *In the set-up of one-way random effects model with unknown mean θ and known variance components $[\sigma_t^2$ and $\sigma_e^2]$, the allocation $A(n, n; 1, 1, \ldots, 1)$ is sufficient for the allocation $A(n, k; n_1, n_2, \ldots, n_k)$ for $1 \leq k \leq n; n_1 + n_2 + \cdots + n_k = n$.*

Proof Note that by definition $A(n, k; n_1, n_2, \ldots, n_k)$ is a combination of k independent allocations of the form $A(n_i, 1, n_i)$. By Lemma 4.5.1, we know that the allocation $A(n_i, n_i; 1, 1, \ldots, 1)$ is sufficient for the allocation $A(n_i, 1; n_i)$ for each $1 \leq i \leq k$. Again a composition of the allocations $A(n_i, n_i; 1, 1, \ldots, 1)$ is equivalent to $A(\sum n_i, \sum n_i; 1, 1, \ldots, 1)$ which is the same as $A(n, n; 1, 1, \ldots, 1)$.

Remark 4.5.1 Some particular cases with $k = 2, 3$ are of special interest.

Theorem 4.5.2 *Of two allocations $A(n, 2; n_1, n_2)$ and $A(n, 2; m_1, m_2)$ subject to $n = n_1 + n_2 = m_1 + m_2$, the former is sufficient for the latter provided $m_1 < n_1 < n_2 < m_2$.*

Proof Not to obscure the essential steps of reasoning, we will go though the following steps.

From the allocation $A(n, 2; n_1, n_2)$, we generate a set of n random observations in two groups.

$$\text{Group} - \text{s} : \; Y(s)_i = \theta + t_s + e(s)_i, 1 \leq i \leq n_s; s = 1, 2.$$

There are two independent (unbiased) estimates for θ, viz. $\bar{y}^{(s)}, s = 1, 2$ with unequal but known variances. Therefore, the combined estimate is formed as usual.

$$\text{(i) } \hat{\theta}^{(s)} = \bar{y}^{(s)},\ V_s = \left[\sigma_{\mathrm{t}}^2 + \sigma_{\mathrm{e}}^2/n_s\right];\ s = 1, 2;$$

$$\text{(ii) } \hat{\theta}^{(c)} = \frac{\bar{y}^{(1)}/V_1 + (\bar{y})^{(2)}/V_2}{1/V_1 + 1/V_2};\ \text{(iii) } V_{\mathrm{c}} = \frac{1}{1/V_1 + 1/V_2} = \frac{V_1 V_2}{V_1 + V_2}.$$

We now make a comparison of $V_{\mathrm{c}}(n_1, n_2)$ for the allocation (n_1, n_2) against $V_{\mathrm{c}}(m_1, m_2)$ for the competing allocation (m_1, m_2) when $n_1 + n_2 = m_1 + m_2 = n;\ m_1 < n_1 < n_2 < m_2$.

Lemma 4.5.2 *In the set-up of one-way random effects model with unknown mean θ and known variance components σ_t^2 and σ_e^2, variance V_c for the combined estimate is smaller for (n_1, n_2) combination compared to the competing (m_1, m_2) combination, uniformly in the variance components σ_t^2 and σ_e^2.*

Proof Without any loss of generality, we can set $\sigma_{\mathrm{e}}^2 = 1$ and then also set $k = \sigma_{\mathrm{t}}^2$. Using the expression for V_{c} in terms of (n_1, n_2) and then again in terms of (m_1, m_2), the comparison amounts to:

$$\frac{(k + 1/n_1)(k + 1/n_2)}{2k + 1/n_1 + 1/n_2} < \frac{(k + 1/m_1)(k + 1/m_2)}{2k + 1/m_1 + 1/m_2}.$$

Upon simplification and equating the coefficients of k^3, k^2, k and the constant term, we obtain an equivalent inequality

$$k^2\left[(1/m_1 + 1/m_2) - (1/n_1 + 1/n_2)\right] + 2k\left[1/m_1 m_2 - 1/n_1 n_2\right] > 0$$

which is easy to validate, since $m_1 < n_1 < n_2 < m_2;\ m_1 + m_2 = n_1 + n_2 = n$.

This result is fundamental and necessary in establishing sufficiency of the allocation $A(n, 2; n_1, n_2)$ for the competing allocation $A(n, 2; m_1, m_2)$ whenever $m_1 < n_1 < n_2 < m_2$), subject to $n = n_1 + n_2 = m_1 + m_2$.

For the allocation $A(n, 2; m_1, m_2)$, let the prospective random observations be, respectively, denoted by $(U_1, U_2, \ldots, U_{m_1})$ and $(V_1, V_2, \ldots, V_{m_2})$. We provide a way of generating the U's and V's starting from the $(Y(s)_i, 1 \le i \le n_s; s = 1, 2)$'s.

Toward this, we generate a set of n linear error functions W_i's following (possibly a singular) n-variate normal distribution and having the known dispersion matrix Σ in the form of a partitioned matrix as indicated below.

$$\Sigma = \left[\mathbf{A}\ \mathbf{B}\ ;\ \mathbf{B}'\ \mathbf{C}\right];\ \mathbf{A} = \sigma_{\mathrm{e}}^2\mathbf{I} + \sigma_{\mathrm{t}}^2\mathbf{J} - V_{\mathrm{c}}\mathbf{J};$$

$$\mathbf{B} = -V_{\mathrm{c}}\mathbf{J};\ \mathbf{C} = \sigma_{\mathrm{e}}^2\mathbf{I} + \sigma_{\mathrm{t}}^2\mathbf{J} - V_{\mathrm{c}}\mathbf{J}.$$

In the above, matrices **I** and **J** are of applicable orders $m_1 \times m_1, m_1 \times m_2, m_2 \times m_2$ leading to **A** of order $m_1 \times m_1$, **B** of order $m_1 \times m_2$ and **C** of order $m_2 \times m_2$.

It follows that $\theta^{(c)} + W_i, 1 \le i \le m_1$ have the same joint distribution as the U's while

$$\theta^{(c)} + W_i, m_1 + 1 \le i \le m_1 + m_2 = n$$

have the same joint distribution as the V's. It must be noted that in the above, we have tacitly assumed that the partitioned matrix Σ as defined, is non-negative definite. We establish this fact below.

Lemma 4.5.3 *In the set-up described above, the partitioned matrix Σ is non-negative definite, irrespective of the values assumed by the variance components σ_t^2 and σ_e^2, whatever be $m_1 < n_1 < n_2 < m_2$; $m_1 + m_2 = n_1 + n_2 = n$.*

Proof The partitioned matrix Σ can be converted into a quadratic form in terms of $(x_1, x_2, \ldots, x_{m_1}) = \mathbf{x}'$ and $(y_1, y_2, \ldots, y_{m_2}) = \mathbf{y}'$ as

$$Q(\mathbf{x}, \mathbf{y}) = (\mathbf{x}', \mathbf{y}')\Sigma(\mathbf{x}', \mathbf{y}')'.$$

Further, let us also set $\mathbf{A} = a\mathbf{I} + b\mathbf{J}, \mathbf{B} = c\mathbf{J}, \mathbf{C} = c\mathbf{I} + d\mathbf{J}$. Upon expansion, the quadratic form can be represented as

$$Q(\mathbf{x}, \mathbf{y}) = \left[a\mathbf{x}'\mathbf{x} + b(\mathbf{x}'\mathbf{1})^2 + 2c(\mathbf{x}'\mathbf{1})(\mathbf{y}'\mathbf{1}) + d(\mathbf{y}'\mathbf{y}) + e(\mathbf{y}'\mathbf{1})^2\right].$$

Since $a = d = \sigma_\mathrm{e}^2 > 0$ and $\mathbf{x}, \mathbf{y}$ are of orders m_1, m_2, respectively, by applying C-S inequality, we derive

$$Q(\mathbf{x}, \mathbf{y}) \ge \left[a(\mathbf{x}'\mathbf{1})^2/m_1 + b(\mathbf{x}'\mathbf{1})^2 + 2c(\mathbf{x}'\mathbf{1})(\mathbf{y}'\mathbf{1}) + d(\mathbf{y}'\mathbf{1})^2/m_2 + e(\mathbf{y}'\mathbf{1})^2\right].$$

This last expression is again a quadratic function $Q(p, q) = (a/m_1 + b)p^2 + 2cpq + (d/m_2 + e)q^2$ where $a = d = \sigma_\mathrm{e}^2 > 0, b = e = \sigma_\mathrm{t}^2 - V_\mathrm{c}, c = -V_\mathrm{c}$.

Let us recall that V_c is based on (n_1, n_2) combination and Lemma 5.2 asserts that this is smaller than the corresponding expression involving (m_1, m_2). We now argue that $a/m_1 + b > 0$.

(i) $a/m_1 + b = \sigma_\mathrm{t}^2 + \sigma_\mathrm{e}^2/m_1 - V_\mathrm{c} = V(m_1) - V_\mathrm{c}$

$$= V(m_1) - \frac{V(n_1)V(n_2)}{V(n_1) + V(n_2)} > V(m_1) - \frac{V(m_1)V(m_2)}{V(m_1) + V(m_2)} > 0.$$

Likewise, $d/m_2 + e > 0$. Further,

$$\begin{aligned}(a/m_1 + b)(d/m_2 + e) - c^2 &= [V(m_1) - V(c)]\,[V(m_2) - V(c)] - V(c)V(c) \\ &= V(m_1)V(m_2) - [V(m_1) + V(m_2)]\,V(c) \\ &= V(m_1)V(m_2)\left[1 - V(c)/V^*\right] > 0,\end{aligned}$$

where

$$V^* = V(m_1)V(m_2)/\,[V(m_1) + V(m_2)] > V(c),$$

as demonstrated in Lemma 4.5.2. This suggests that the quadratic function $Q(p, q)$ is non-negative whatever be the real numbers p and q. The proof of Theorem 5.2 is thus completed.

We may now deduce several results on comparison of allocations.

Corollary 4.5.1 *For* $n = 2m + 1$, *the allocation* $A(n = 2m + 1, k = 2; n_1 = m, n_2 = m + 1)$ *is sufficient for any other competing allocation.*

By repeated application of Corollary 4.5.1, we may deduce further results.

Corollary 4.5.2 *For* $n = 3m + 1$, *the allocation* $A(n = 3m + 1, k = 3, n_1 = n_2 = m, n_3 = m + 1)$ *is sufficient for any other competing allocation.*

Corollary 4.5.3 *For* $n = 3m + 2$, *the allocation* $A(n = 3m + 2, k = 3, n_1 = m, n_2 = n_3 = m + 1)$ *is sufficient for any other competing allocation.*

4.6 Conclusion

Much of what have been discussed in this chapter are the results of comparatively recent studies carried out by one of the authors in the form of collaborative research. We believe that the readers will find the topics and derivations of results quite appealing and thought-provoking. For the Sects. 4.3 and 4.4, we refer to Dhungana et al. (2010), Tiensuwan et al. (2010) and Tiensuwan and Dhungana (2011). Further, we refer to DeGroot (1966), Sinha (1971) and Stepniak (1982) for the material in Sect. 4.5. The essential contents of Sect. 4.2 are fairly standard and are to be found in text books.

References

Blackwell, D. (1951). Comparison of experiments. In *Proceedings of the Second Berkeley Symposium on Mathematical Statistics and Probability* (pp. 93–102). University of California Press.

Blackwell, D. (1953). Equivalent comparisons of experiments. *Annals of Mathematical Statistics, 24*, 265–272.

DeGroot, M. H. (1966). Optimal allocation of observations. *Annals of the Institute of Statistical Mathematics, 18*, 13–28.

Dhungana, A. R., Tiensuwan, M., & Sinha, B. K. (2010). Sufficiency in bivariate and trivariate normal populations. *Far East Journal of Theoretical Statistics, 32*(1), 59–80.

Sinha, B. K. (1971). Comparison of some experiments from sufficiency consideration. *Annals of the Institute of Statistical Mathematics, 25*, 501–520.

Stepniak, C. (1982). Optimal allocation of observations in one way random normal model. *Annals of the Institute of Statistical Mathematics, 34*(Part A), 175–180.

Tiensuwan, M., Dhungana, A. R., & Sinha, B. K. (2010). Sufficiency in linear and quadratic regression models. *Journal of Statistical Theory and Applications, 9*(3), 387–404.

Tiensuwan, M., & Dhungana, A. R. (2011). Understanding sufficiency in one-way random effects model. *Sankhya B, 73*, 263–275.

Chapter 5
Estimation of the Size of a Finite Population with Special Features

Learning Objectives
This is again a specific problem formulation which is most likely to be innovative, challenging, and worth pursuing. We engage the readers to critical thinking right from the beginning. Here we deal with a finite population of [ultimate] units [individuals] without any direct access to any of them to start with. Access is achieved only through a 'network' involving what are termed as Reference Units [which are finite in number and identifiable]. The subject matter is developed slowly with reference to [illustrative] example and the problem of estimation of the population size involving ultimate units is posed. The set-up calls for new techniques of unbiased/biased estimation. We enjoyed working on this formulation and developing suitable analytical results which we share with the readers with utmost pleasure. We fondly hope serious researchers will find the developments stimulating and worth pursuing.

5.1 Introduction and Literature Review

Estimation of the unknown size (N) of a finite population [in different contexts/application areas] has been of interest to the statisticians. The well-known 'Catch-Recatch' method is talked about frequently. There are available non-trivial and often-needed alternatives to this method as well. We will skip any discussions on the standard Catch-Recatch methodology altogether. Instead, a few references are cited at the end. It may be noted that there are other equivalent versions of this method, viz. Catch–Mark–Release–Recatch method.

In this chapter, we will deal with finite populations having peculiar/special features and discuss about the problem of estimation of the unknown size (N) of such a population. The peculiarity of the populations lies in the fact that none of the population units of our interest are directly accessible to the investigator. Instead, there is a well-defined finite population of what are called Reference Units [RUs] and these RUs may be accessed directly. As against this, the former population [of our primary interest] is referred to as that of the Ultimate Units [UUs]. Further to this, there is

© The Author(s), under exclusive license to Springer Nature Singapore Pte Ltd. 2024

M. Pal and B. K. Sinha, *Selected Topics in Statistical Inference*,
https://doi.org/10.1007/978-981-97-2592-2_5

a well-defined *network* connecting the population of the RUs to that of the UUs. A sample of RUs will naturally create a partial view of the population network. Based on this sample network, it is required to unbiasedly estimate the size of the population of UUs. We review the literature [which is scanty anyway] and discuss some thoughts toward satisfactory acceptable solutions to this fascinating and seemingly 'trivial' problem.

We refer to Seber (1982) for literature review and a host of results in standard formulations of this problem. Note, however, that we are dealing with a non-standard set-up and we are following a network approach to reach the ultimate units of our interest.

The first non-trivial study of the problem of estimation of N based on a partial view of the network, in terms of exposure to some m out of M Reference Units [drawn according to SRSWOR(M, m) sampling procedure], seems to be due to Kiranandan (1976). Following this, there are two more studies. Maiti et al. (1992) derived interesting results on unbiased estimation of N, being completely unaware of Kiranandan's work. Later, Sinha et al. (2006) took cognizance of Kiranandan's work and presented an equivalent version, along with a critical appraisal of K's work. These studies address the above problem with/without additional information [which may be made available in terms of what has been termed as 'probing']. After a random sample [$s(m)$] of m Reference Units has been selected and the sample network has been exhibited, we have a count of the number of Ultimate Units so captured [say, $n(s(m))$] and their individual incidences within the sampled Reference Units. In most studies, this alone would constitute our data and we should proceed toward estimation of N from here. This corresponds to the scenario of 'no probing'. It seems to be an utterly impossible proposition to go anywhere from here ! Kiranandan (1976) succeeded in deriving a novel formula for $\hat{N}$, an unbiased estimator for N, in terms of the frequency counts of the Ultimate Units so captured [which are $n(s(m))$ in number] via the sample network based on $n(s(m))$ UUs and $s(m)$ RUs. However, unfortunately in most applications, $\hat{N}$—so computed—may turn out to be a large negative quantity ! There are other undesirable features of $\hat{N}$ as well. It may turn out to be less than the actual number of Ultimate Units captured/observed in the process of sampling. The criterion of unbiasedness thus seems to take its toll in this process. One of the objectives in the present study is to alert the readers about these undesirable features of $\hat{N}$ and create a platform for discussion on other *reasonable* estimators for N, doing away with unbiasedness, if needed, in the process. Toward this, we will only mention one recent study (Padmawar & Sinha 2016).

We must hasten to add that, unlike the contents of all other chapters, the results presented and discussed in this chapter are likely to be totally new to the readers at large. We fervently hope that the readers will get a feel for 'research' in a totally new area and will be enlightened by our presentations and discussions.

5.2 Special Features of Finite Populations

Estimation of N, the unknown size of a finite population, is an interesting topic by itself and we get exposure to the celebrated 'Catch-Recatch' technique in an elementary course in inference. However, there is much more to be told, studied, and researched. This is our topic in this chapter.

We will slowly develop the problem formulation and the techniques toward its solution. We intend to deal with human populations, but with special features.

We consider a general population and within it, there are individuals (i) having access to food coupons only through distribution centers or service centers, or, (ii) suffering from a rare disease and necessarily visiting one or more diagnostic centers for treatments, and the like. These individuals are otherwise not identifiable in the general population. The only way to identify them is by visiting the appropriate centers meant for such purpose. Below we consider the case of food coupon beneficiaries and service centers. Once more, we reiterate that our purpose is to estimate the unknown number of beneficiaries in the entire population.

5.2.1 Concepts of Reference Units and Ultimate Units

Let there be M service centers catering to a total of N distinct individual beneficiaries. An individual beneficiary may, however, avail of services from multiple service centers, without any restriction whatsoever ! To start with, it is stipulated that the individual beneficiaries are neither directly accessible nor are they identifiable. The only way to have access to them is through visiting one or more of the service centers to which they are affiliated/attached/registered. The M service centers are distinct and identifiable and form a finite labeled population of what may be termed as **Reference Units**—abbreviated as RUs. We assume this number M to be known. In case we decide to check/inspect all the service centers, we gain access to all the N beneficiaries through the underlying network which forms a bipartite di-graph, in the language of graph theory. On the other hand, in case M is large so that complete enumeration is cost-prohibitive, we may as well take recourse to sampling of the service centers and end up with an incomplete description of the underlying network involving a part of the population of beneficiaries. Then the question of unbiased estimation of the total size N, i.e., total number of beneficiaries, is a non-trivial exercise. The beneficiaries may be said to form a population of **Ultimate Units**—abbreviated as UUs. The literature on estimation of the size of the population of UUs is scanty, and the results available are not very convincing/useful. Our purpose is to provide a background to make a case for additional thoughts in this fascinating area of research.

5.2.2 Concepts of a Bipartite Graph and a Network of Reference Units and Ultimate Units

Imagine that a sample of RUs of size m has been selected according to SRSWOR (M, m). [Note that this could as well be any other Fixed Size (m), i.e., FS(m) sampling scheme with desirable properties]. Since, apriori, the UUs are neither identifiable nor bear any labeling for the total count N [which is itself unknown], we may assign 'labeling' of these UUs as follows.

Set $\text{RU}_{s(m)} = [(\text{RU}_{i_1}, \text{RU}_{i_2}, \ldots, \text{RU}_{i_m}); 1 \leq i_1 < i_2 < \cdots, < i_m \leq M]$. Note that the RUs are already labeled as $(\text{RU}_1, \text{RU}_2, \ldots, \text{RU}_M)$ in some convenient order and we may conveniently/deliberately simplify the notations by dropping the prefix RU. WOLG, let us set $s(m) = [1, 2, \ldots, m]$. We pick up the first RU, i.e., RU_1 and find out a listing of UUs contained therein. If this number is n_1, we assign serial numbers $\text{UU}_1, \text{UU}_2, \ldots, \text{UU}_{n_1}$ to these n_1 UUs in some manner. There is no hard and fast rule for doing so. Next we check the sampled second RU, i.e., RU_2. If there are some common UUs with RU_1, we simply carry out the identification and do not assign any extra label to such UUs. If there are 'new' UUs, we assign them serial numbers starting from $\text{UU}_{n_1} + 1$ and so on until these new, i.e., not covered by RU_1 UUs are exhausted. We continue like this till we end up with the last sampled RU, i.e., RU_m. Over here, we list all those UUs, if any, which have not so far been covered by the union of the RUs RU_1 to RU_{m-1} previously taken up for listing of UUs. Incidentally, in the process, we have generated a **Bipartite Graph** involving the two sets : RUs and UUs. Further, while the selected number of RUs is fixed [in this case, it is m out of M], the number of UUs captured in the process is a random quantity. We have previously denoted this number by $n(s(m))$. Henceforth, we may use the notation n and the resulting sample of UUs may be denoted by UUs(n). To distinguish between $s(m)$ and $s(n)$, we may use the notations RUs(m) and UUs(n), respectively. We must note that the bipartite graph so generated is based on the sampled RUs and the *induced* sample of UUs.

5.2.3 Sampling from Reference Units and Creation of a Sample Network

In the above, we have mentioned about the population of M RUs and selection of a fixed size of m RUs, adopting SRSWOR or any other convenient FS(m) sampling design. Based on this random selection of m RUs, conceptually we are capable of generating an underlying network—connecting the sampled RUs and the *induced* UUs. This network will be referred to as a **Sample Network**. There is an underlying population network which is **not** accessible to us *unless* we carry out complete enumeration of all the M RUs.

Table 5.1 Population network involving $M = 10$ reference units

Reference units	Ultimate units
1	[(1, 2, 3)]
2	[(1); (4, 5, 6, 7)]
3	[(2, 6); (8, 9, 10, 11, 12)]
4	[(5, 10); (13, 14)]
5	[(3, 7, 9, 13); (15, 16, 17, 18, 19)]
6	[(5, 12, 17, 19); (20, 21, 22, 23, 24)]
7	[(8, 13, 18, 20, 23, 24); (25, 26, 27)]
8	[(10, 15, 20, 23, 25, 27); (28, 29, 30, 31)]
9	[(2, 6, 8, 12, 15, 21, 28); (32, 33, 34, 35)]
10	[(5, 8, 10, 13, 15, 17, 21, 29); (36, 37, 38)]

Before proceeding further, we will take up an illustrative example. Consider $M = 10$ Service Centers [SCs], i.e., RUs with reference to a given population in respect of providing Food Coupons. A beneficiary has free access to any one or two or even more SCs without any restriction whatsoever. We will be using RUs and SCs interchangeably and conveniently in this example dealing with Food Coupons.

Ideally, if we are in a position to scan through the records of all the $M = 10$ SCs, we have a perfect idea about the entire and complete Network of ALL SCs crossed with ALL Beneficiaries in the reference population. This is the Population Network we are referring to. In that case, there is no need for any further statistical analysis at sampling level. However, most often, it is prohibitive to carry out a thorough investigation at the population level involving ALL the $M = 10$ SCs. That is why we suggest that one may adopt a convenient $FS(m)$ sampling scheme.

To fix the ideas, at first we display the Population Network. One such [fictitious] population network is given in Table 5.1.

The entries in Table 5.1 are self-explanatory numbers which represent the serial numbers attached to the UUs. To create the entries in this table, at first the RUs have been serially numbered from 1 to 10. Then we check with RU_1. It turns out that there are 3 UUs attached to it. Those three UUs are numbered as UU_1, UU_2, UU_3. Next we check with RU_2. There are 5 UUs attached to it. Of these, 1 is 'old' and the rest are 'new'. These new UUs are now serially numbered as UU_4, UU_5, UU_6, UU_7 and are thus separated from the old serial numbers. Old and New UUs are separately placed inside brackets to make the findings self-complete and self-explanatory. At the end, we find that altogether there are 38 UUs in this collection. It must be noted that until and unless we check through the entire collection of the RUs, we do not have a complete picture of the incidence patterns of the UUs. It is an instructional exercise for the students to start with a different arrangement of the RUs, say [2, 7, 10, 4, 1, 9, 3, 6, 8, 5] and reconstruct the incidence pattern. It will be observed that the final count [of 38 UUs] is invariant with respect to rearrangement of the RUs.

The reader can now visualize the bipartite di-graph. To the left, we draw a vertical bar representing the RUs in their natural order and to the right we draw the same for the UUs. A 'directed' arrow from RU_i to UU_j simply means that the RU_i contains the UU_j. The 'Incidence Matrix' underlying the above data is given in Table 5.2. Incidentally, at the end, we also have information on R_j which is the number of RUs resulting in UU_j in the entire network. For example, $R_1 = 2$, $R_{10} = 4$,etc. In Table 5.2, we also display these R_j-values for all UUs.

We note that complete enumeration of all the RUs will provide a complete picture of the network and hence the total number of UUs will be ascertained and we will know much more than that. So far—except for complexity of the underlying network—we have perfect knowledge about the movements of the UUs across the RUs.

In reality, however, it becomes cost-prohibitive to carry out a complete enumeration of the entire collection of M RUs for any reasonable value of M. It is interesting to note that the entire collection of RUs is available to the statistician and because these are all distinct, one can label them in any arbitrary manner and come up with $RU_1, RU_2, \ldots, RU_M$— thereby arriving at a finite labeled population of M RUs. It is now natural for the statistician to take recourse to a convenient sampling scheme, say $FS(m)$ sampling scheme with desirable properties of the first and second-order inclusion probabilities. Vide Hedayat and Sinha (1991). The statistician is thus naturally drawn to the techniques of finite population sampling and inference. As a particular case, SRSWOR(M, m) becomes a natural choice.

We now go back to the example of the population network in Table 5.1 and consider the sampling aspect. We choose $m = 3$ and suppose SRSWOR(10, 3) results in the random selection of (RU_3, RU_5, RU_9). How would the data collection effort take shape now under sampling of 3 RUs ? What would be the resulting sample network of RUs versus [induced] UUs ? It is a good instructional lesson for all of us.

To construct the sample network, we will scan through these 3 selected RUs in their natural order, i.e., as RU_3, RU_5, RU_9 in this order. Referring to Table 5.1, we find that RU_3 contains 7 UUs which are now denoted as (UUs_1, UUs_2, UUs_3, UUs_4, UUs_5, UUs_6, UUs_7). In later contexts, it will be helpful to distinguish between UUP_j and UUs_j—the UU holding the jth serial number in the population or in the sample. Next we examine RU_5 which is seen to contain 9 UUs of which exactly one, viz. UUs_4 is 'old' and the rest are all 'new'. These new UUs are altogether 8 in number and we label them as (UUs_8, UUs_9, ..., UUs_{15}). Lastly, we have selected RU_9 which resulted in a total of 11 UUs. Of these, 4 UUs have intersection with those in RU_3 [viz. UUs_1, UUs_2, UUs_3, UUs_7] and only one, viz. SSu_{11} is in common with RU_5. So we still have 6 new UUs. These are denoted by (UUs_{16}, UUs_{17}, ..., UUs_{21}). Hence in this sample of 3 RUs, we have captured a set of 21 UUs.

Expressed in an equivalent manner, in between RU_3 and RU_5, we have an intersection of only one UU [i.e., UUP_9] which is the same as UUs_4 ; likewise, between RU_3 and RU_9, there are 4 UUs common [viz. UUP_2, UUP_6, UUP_8, UUP_{12}] [having equivalent representation as UUs_1, UUs_2, UUs_3, UUs_7] and finally, between RU_5 and RU_9, there is again exactly one UU common, i.e., UUP_{15}, or UUs_{11}.

Table 5.2 Incidence matrix of the population network of 10 RUs and 38 UUs

Reference Units	1	2	3	4	5	6	7	8	9	10
1	1	1	1	0	0	0	0	0	0	0
2	1	0	0	1	1	1	1	0	0	0
3	0	1	0	0	0	1	0	1	1	1
4	0	0	0	0	1	0	0	0	0	1
5	0	0	1	0	0	0	1	0	1	0
6	0	0	0	0	1	0	0	0	0	0
7	0	0	0	0	0	0	0	1	0	0
8	0	0	0	0	0	0	0	0	0	1
9	0	1	0	0	0	1	0	1	0	0
10	0	0	0	0	1	0	0	1	0	1
R_j	2	3	2	1	4	3	2	4	2	4
Reference Units	11	12	13	14	15	16	17	18	19	20
1	0	0	0	0	0	0	0	0	0	0
2	0	0	0	0	0	0	0	0	0	0
3	1	1	0	0	0	0	0	0	0	0
4	0	0	1	1	0	0	0	0	0	0
5	0	0	1	0	1	1	1	1	1	0
6	0	1	0	0	0	0	1	0	1	1
7	0	0	1	0	0	0	0	1	0	1
8	0	0	0	0	1	0	0	0	0	1
9	0	1	0	0	1	0	0	0	0	0
10	0	0	1	0	1	0	1	0	0	0
R_j	1	3	4	1	4	1	3	2	2	3
Reference Units	21	22	23	24	25	26	27	28	29	30
1	0	0	0	0	0	0	0	0	0	0
2	0	0	0	0	0	0	0	0	0	0
3	0	0	0	0	0	0	0	0	0	0
4	0	0	0	0	0	0	0	0	0	0
5	0	0	0	0	0	0	0	0	0	0
6	1	1	1	1	0	0	0	0	0	0
7	0	0	1	1	1	1	1	0	0	0
8	0	0	1	0	1	0	1	1	1	1
9	1	0	0	0	0	0	0	1	0	0
10	1	0	0	0	0	0	0	0	1	0
R_j	3	1	3	2	2	1	2	2	2	1
Reference Units	31	32	33	34	35	36	37	38		
1	0	0	0	0	0	0	0	0		
2	0	0	0	0	0	0	0	0		
3	0	0	0	0	0	0	0	0		
4	0	0	0	0	0	0	0	0		
5	0	0	0	0	0	0	0	0		
6	0	0	0	0	0	0	0	0		
7	0	0	0	0	0	0	0	0		
8	1	0	0	0	0	0	0	0		
9	0	1	1	1	1	0	0	0		
10	0	0	0	0	0	1	1	1		
R_j	1	1	1	1	1	1	1	1		

Presumably this is our data. At this stage, are we in a position to develop adequate and appropriate theory for unbiased estimation of N, the total number of UUs in the population ? Even under SRSWOR(10, 3) sampling scheme, this seems to be an impossible proposition—mainly because we are looking for an *unbiased* estimator for N. Incidentally, the interested readers can easily draw the sample bipartite digraph connecting 3 RUs with the 21 UUs. Further to this, interestingly, we find that we also have data for each captured UUs [through our sampling effort] on the number of RUs [out of the sampled 3 RUs] on which it is 'incident'. Naturally, each such incidence number is at least one and does not exceed three. From the population network, we had exact information on R_j for each UU_j. The sample counter part may be denoted by r_j. It is better to attach indicators to the labels of the UUs since only sampled UUs will possess such information on r's. Thus, for example, $I(\text{RU}_{s(3)}; (j))$ would provide necessary information where the indicator value is 1 iff UUs_j is contained in the collection of UUs based on the sample $\text{RU}_{s(3)}$. Accordingly, we note that $\sum_j I(\text{RU}_{s(3)}; (j)) = r_j$. We hope we have made the notations clear and meaningful.

Interestingly, this impossible proposition [of suggesting an unbiased estimator for N] was made possible by Kiranandan (1976) in his Ph.D. Dissertation at Harvard University. We will take up a detailed presentation of this study in Sect. 5.3.2.

Over there, we will closely follow the work done by Sinha et al. (2006). Prior to this, we would like to take up another [relatively easier !] approcah for unbiased estimation of N. Here we will closely follow the work of Maiti et al. (1992) and this is contained in Sect. 5.3.1.

Both the estimation problems are discussed in Sect. 5.3.

5.3 Unbiased Estimation of the Size of a Specially Featured Population

5.3.1 Concepts of Probing and No-Probing

Let us look at the sample incidence matrix and the resulting sample quantities r_js. Is it possible to entertain the idea of 'probing' ? By that we mean we can approach all or a subset of these sampled UUs with the 'proposal / request' to release information on the number of *additional* [unsampled] RUs containing the specific UU whom we are requesting ! Now that we have hypothetically accsess to the population network, if we approach the sampled UUs and they oblige us by providing/releasing the requested information, then we will be in a position to assess the correctness of the figures and prepare an extended table as follows.

In Table 5.4, the notations are supposed to be self-explanatory. After probing, information on Total Count of RUs in the entire population for each of the induced UUs becomes available. Before probing, this count [for every incident UU] was based on the evidence provided by Sampled RUs. On the other hand, after probing,

Table 5.3 Sample of UUs induced by selected RUs under SRSWOR($M = 10, m = 3$)

Selected RUs	Induced UUs
RU_3	[(1, 2, 3, 4, 5, 6, 7)]
RU_5	[(4; 8, 9, 10, 11, 12, 13, 14, 15)]
RU_9	[(1, 2, 3, 7, 11; 16, 17, 18, 19, 20, 21)]

Table 5.4 Induced UUs and their incidence counts

Induced UUs	Incidence counts among sampled RUs	Net effect after probing
UUs_1	$r_1 = 2$	$R_1 = 2 + 1 = 3$
UUs_2	$r_2 = 2$	$R_2 = 2 + 1 = 3$
UUs_3	$r_3 = 2$	$R_3 = 2 + 3 = 5$
UUs_4	$r_4 = 2$	$R_4 = 2 + 0 = 2$
UUs_5	$r_5 = 1$	$R_5 = 1 + 3 = 4$
UUs_6	$r_6 = 1$	$R_6 = 1 + 0 = 1$
UUs_7	$r_7 = 2$	$R_7 = 2 + 1 = 3$
UUs_8	$r_8 = 1$	$R_8 = 1 + 1 = 2$
UUs_9	$r_9 = 1$	$R_9 = 1 + 1 = 2$
UUs_{10}	$r_{10} = 1$	$R_{10} = 1 + 3 = 4$
UUs_{11}	$r_{11} = 2$	$R_{11} = 2 + 2 = 4$
UUs_{12}	$r_{12} = 1$	$R_{12} = 1 + 0 = 1$
UUs_{13}	$r_{13} = 1$	$R_{13} = 1 + 2 = 3$
UUs_{14}	$r_{14} = 1$	$R_{14} = 1 + 1 = 2$
UUs_{15}	$r_{15} = 1$	$R_{15} = 1 + 1 = 2$
UUs_{16}	$r_{16} = 1$	$R_{16} = 1 + 2 = 3$
UUs_{17}	$r_{17} = 1$	$R_{17} = 1 + 1 = 2$
UUs_{18}	$r_{18} = 1$	$R_{18} = 1 + 0 = 1$
UUs_{19}	$r_{19} = 1$	$R_{19} = 1 + 0 = 1$
UUs_{20}	$r_{20} = 1$	$R_{20} = 1 + 0 = 1$
UUs_{21}	$r_{21} = 1$	$R_{21} = 1 + 0 = 1$

this count reflects the total count of such incidences in the entire collection of M RUs. However, it is only for the induced UUs in the sample.

In passing, we may note that on final count, information accrued through probing amounts to Y_j for the sampled UUs_j in the framework of unified theory of sampling. Maiti et al. (1992) followed this approach and used R_j for Y_j and applied sampling techniques to arrive at an unbiased estimate of N, the total number of Ultimate Units in the entire population, based on data under/after probing.

Once more we fix the notations.

We have denoted by M the total number of RUs in the population and by m the sample size for selection of RUs under random sampling. We have denoted by N

the total [unknown] number of Ultimate Units in the population. Let us denote by n the sample number of UUs captured through our sampling effort. For the UUs_j in the sample, we have denoted by r_j the number of sampled RUs connected to it before probing. We may rightly denote the corresponding count after probing as R_j. Clearly, for each UUs_j in the sample, $1 \le r_j \le R_j \le M$. These are essentially all the information available to us.

Note that we have assumed that each of the n UUs in our induced sample has been asked / requested to adopt probing and extend the count r_j to R_j.

5.3.2 Estimation of N Under Probing: Theory and Examples

When SRSWOR(M, m) is adopted for selection of m RUs out of M RUs, Maiti et al. (1992) suggested the formula

$$\hat{N} = (M/m)\sum_j r_j/R_j$$

and it turns out that this estimator is unbiased. In the particular case of $m = 1$, it readily follows that $\hat{N} \ge n$ where n is the sample count of induced UUs.

For the cases of $m > 1$, the estimator given above is unbiased; however, it is not clear if $\hat{N} \ge n$. based on data for each of the all possible M_{c_m} samples of RUs. At this stage, we indicate a proof of unbiasedness of $\hat{N}$—first for $m = 1$ and then for $m > 1$.

Note that when $m = 1$, we are randomly selecting just one RU out of M RUs. Hence the UUs can be captured through the use of natural Indicator Functions, i.e., by $I(j|i) = 1$ if UU_j is contained in the set of UUs induced by RU_i. In other words, each r_j is simply 1 or 0. Hence,

$$\sum_i I(j|i) = R_j; \quad j = 1, 2, \ldots.$$

This yields

$$E[\hat{N}] = M \times \sum_j \left[\sum_i I(j|i)/MR_j\right] = \sum_j I(j) = N,$$

where $I(j) = 1$ for every UU_j in the entire network. Logically, therefore, this last expression yields the parameter N to be estimated.

Maiti et al. (1992) started by considering a binary matrix of order $M \times N$ where M stands for the known number of Service Centers and N is the unknown number of Beneficiaries. As the set-up suggests, the entries in the matrix are 0s and 1s. That is what we have considered here. In Sinha et al. (2006), the entries have been generalized to any set of real values.

In the general case, it follows that whenever $I(j) = 1$ for some UU_j in the population network, under SRSWOR(M, m) sampling,

$$E[r_j) = mR_j/M,$$

R_j being the total count of RUs in the population network having UU_j incident on each one of them. To avoid triviality, note that $R_j = 0$ simply means that $I(j) = 0$ outright and hence this specific UU does not exist in the existing framework of Reference Units versus Ultimate Units.

Note further that this estimation is possible since, under probing, R_j becomes available whenever $I(j) = 1$ and UU_j is included in the sample as UUs_j so that the investigator gets hold of R_j. Strictly speaking, under SRSWOR(M, m) sampling, for every UU_j for which $I(j) = 1$, r_j has Hypergeometric Distribution with parameters $[M, R_j; m]$ and, hence, $[0 \leq r_j \leq min\ (R_j, m)]$. This explains the validity of the formula for $\hat{N}$ in general for $m \geq 1$.

We now take an illustrative example, and for the purpose of illustration of the computations, we may use the same network as shown above.

Example 5.3.1 In the given network, we have $M = 10, m = 3$. Further, while RU_3, RU_5, RU_9 are the selected RUs, these collectively induct a total of 21 UUs. Moreover, with probing applied to each of these 21 inducted UUs, the frequency counts [viz. both r_j and R_j] are given in Table 5.3. From this, we derive

$$\sum_j r_j/R_j = [(2/2) \times 1 + (2/3) \times 3 + (2/4) \times 1 + (2/5) \times 1] \\ + [(1/1) \times 6 + (1/2) \times 5 + (1/3) \times (2) + (1/4) \times 2],$$

and this simplifies to $[3.9 + 9.6667 = 13.5667]$.

Therefore, on final count,

$$\hat{N} = (M/m) \times \left[\sum_j r_j/R_j\right] = (10/3) \times 13.5667 = 45.222 = 46 \text{ [approx.]}.$$

Remark 5.3.1 In case an arbitrary $FS(m)$ sampling design is adopted, the formula changes to

$$N(s(R)) = \sum_i I(i, s(R)) \sum_j I(j|i)/R_j \pi_i.$$

Here $I(i, s(R)) = 1(0)$ when RU_i is (is not) in the sample $s(R))$ of RUs. Likewise, $I(j|i)$ is the usual indicator relating to RU_i and UU_j in the incidence network. Further, with reference to the adopted $FS(m)$ sampling design, π's are the first-order inclusion probabilities of the RUs.

The above formula was given in Maiti et al. (1992). Available multiple estimators due to Sirken (1970), Birnbaum and Sirken (1965) are of the above type.

With this, we close our discussion on unbiased estimation of N under probing.

5.3.3 Estimation of N Without Probing: Theory and Examples

This time, as before, we start with a random selection of RUs and arrrive at the induced collection of UUs. We do not feel encouraged to undertake the task of probing—not even for a subset of the induced UUs. We go back to the example considered in the earlier set-up. Thus there are altogether 10 RUs and we have selected 3 of them and the data accrued on the induced UUs results in 21 of them with frequency counts (r_js) as given in Table 5.3. We reproduce the r_j-values here.

$$(2, 2, 2, 2, 1, 1, 2, 1, 1, 1, 2, 1, 1, 1, 1, 1, 1, 1, 1, 1, 1).$$

Without any probing, our data collection effort ends here. It seems quite a daunting task to come up with any clue toward *unbiased* estimation of the total number of UUs, i.e., of N. Kiranandan (1976) did that seemingly impossible task and arrived at a fundamental identity showing a representation of N.

Heuristically, $N = \sum_j I(j)$ where $I(j)$ assumes the value 1 whenever UU_j is a valid entity in the collection of UUs. But the verification comes only through the RUs which are M in number. We may set $I(j|i) = 1$ if and only if UU_j is induced by RU_i. Moreover, $I(j|i)I(j|k) = 1$ if and only if both the indicators assume value 1, and so on. We may now state below one useful identity:

$$I(j) = \sum_{i} I(j|i) - \sum\sum_{i<k} I(j|i)I(j|k) + \sum\sum\sum_{i<k<t} I(j|i)I(j|k)I(j|t) - \cdots + (-1)^{M-1} I(j|1)I(j|2)\ldots I(j|M)$$

The proof of this identity is very elementary. It follows from the counting argument and use of inclusion-exclusion principle. Imagine UU_j is inducted by some t RUs and WOLG, we may assume : $I(j|1) = I(j|2) = \cdots = I(j|t) = 1$. Then the LHS is unity and the RHS is

$$f(t) = t - t_{c_2} + t_{c_3} - t_{c_4} + \cdots$$

It is easy to argue that $f(t) = 1$ for each $t = 1, 2, \ldots$.

We may now state the Fundamental Identity or, the Fundamental Representation Theorem:

Theorem 5.3.1

$$N = \sum_j I(j) = \sum_j \sum_i I(j|i) - \sum_j \sum_{i<k} \sum I(j|i)I(j|k)$$
$$+ \sum_j \sum \sum_{i<k<t} \sum I(j|i)I(j|k)I(j|t) - \cdots$$
$$+ (-1)^{M-1} \sum_j I(j|1)I(j|2) \ldots I(j|M)$$
$$= \sum_i T_i - \sum_{i<k} \sum T_{i,k} + \sum \sum_{i<k<t} \sum T_{i,k,t} + \cdots,$$

where T_i = Number of UUs induced by RU_i, $T_{i,k}$ = Number of UUs simultaneously induced by both RU_i and RU_k, and so on.

Proof It is a matter of summing over all j for which $I(j) = 1$. The rest follows upon simplification.

Remark 5.3.2 To find an unbiased estimator for N, we now go for term by term estimation. If the sample size is $m < M$, term-by-term estimation will terminate after m terms. So we need a condition on the nature of the population network : No Ultimate unit is incident on more than m Reference Units. Our choice of the sample size m for random selection of RUs must be determined by this pre-condition. In applications, this is not hard to meet with by an appropriate choice of m. This point is already mentioned above. Here we have re-emphasized it.

Those of us familiar with the unified theory of sampling and inference will find it a routine exercise to go for term-by-term unbiased estimation by using what are called 'joint inclusion probabilities'. When dealing with a term involving exactly one RU, say RU_i, we will consider dividing by π_i; for two RUs, say RU_i and RU_k, we will divide by $\pi_{i,k}$ and so on. So technically, we also need one necessary condition and this is to ensure positivity of all joint inclusion probabilities, upto and including the mth order ! Note that sampling is taking place invloving only the RUs and hence all these joint inclusion probabilities refer to the M RUs only. In order that all m-ples involving the RUs do possess positive joint inclusion probabilities, it is usually assumed that we are undertaking SRSWOR(M, m) sampling scheme on the population of M RUs ! Of course, there are many other choices and also we have varying probability sampling schemes, such as Midzuno Sampling Schemes, for example. We refer to Hedayat and Sinha (1991) for this aspect of sampling and inference.

Under SRSWOR(M, m) sampling scheme, we know

$$\pi_i = m/M; \pi_{i,k} = m(m-1)/M(M-1) = m^{(2)}/M^{(2)};$$

$$\pi_{i,k,t} = m(m-1)(m-2)/M(M-1)(M-2) = m^{(3)}/M^{(3)}; \ldots.$$

Under SRSWOR(M, m) sampling scheme, an expression for $\hat{N}$ can now be deduced slowly and carefully. We refer to Sinha et al. (2006) for this task and simply reproduce the final expression here.

$$\hat{N} = \sum_f n(f)A(f); \quad A(f) = 1 - (-1)^f[(M-m+f-1)!(m-f)!]/[(M-m-1)!m!], \quad f = 1, 2, \ldots,$$

where $n(f)$ = number of UUs, each with incidence number f in the sample network on m RUs and the induced UUs. The underlying simplification is not at all obvious. The students/researchers have to work around the formula quite carefully and with much attention.

Example 5.3.2 In the example that we have considered above, we have $M = 10, m = 3$ and $[RU_3, RU_5, RU_9]$ are the sampled RUs and these altogether lead to a total of 21 induced UUs. Their frequencies are again reproduced below.

$$(2, 2, 2, 2, 1, 1, 2, 1, 1, 1, 2, 1, 1, 1, 1, 1, 1, 1, 1, 1, 1).$$

We find

$$f = 1, n(1) = 15, A(1) = 1 + 7!/6!3! = 1 + 7/6 = 13/6;$$

$$f = 2, n(2) = 6, A(2) = 1 - 8!/6!3! = 1 - 28/3 = -25/3.$$

Therefore,

$$\hat{N} = \sum_f n(f)A(f) = 15 \times 13/6 - 6 \times 25/3 = 32.5 - 75 = -42.5!!!$$

What's going on ?

Remark 5.3.3 We had alerted the readers about such totally unacceptable estimators for the unknown size of a finite population of UUs. There are primarily two reasons for this. We are taking recourse to indirect, viz. network sampling via the selection and use of RUs. And, further, we require the estimator to be unbiased. Padmawar and Sinha (2016) have made a systematic and detailed study of such consequences. They considered an example of a population network involving $M = 6, m = 3$) and computed numerical values of all possible estimators of N based on the totality of 20 possible samples, each of size 3 under SRSWOR(6, 3) sampling. They observed that though the estimator is unbiased, 6 of the estimators were low/abnormally low. As against the actual value of $N = 10$, there are estimates as low as $1, 2$! Here we ended up with a negative value !!

These are realities while we are dealing with unbiased estimation of N. That is why, Padmawar and Sinha (2016) made a systematic study of the problem in considerable details - with the whole objective of generating meaningful estimates.

In the next section, we will briefly mention about that study based on heuristics.

5.4 Estimation of N: Heuristic Principle

We will briefly discuss one of the two principles suggested in Padmawar and Sinha (2016). For this, once the sample of RUs is selected and the underlying network of UUs is achieved, we stop and study this network carefully and objectively—without any probing efforts.

Padmawar and Sinha (2016) developed a 'Heuristic Method' for extracting information on 'New Entrants [UUs]' from a given set of data through 'Extension Principle'. In this section, we will elaborate on this method.

Our purpose is to arrive at a 'reasonable estimator' of N, starting with a random sample of m out of M RUs. First of all, we will treat the given sample $s(m)$ of m RUs as forming a subpopulation in its own right. The notation $s(m)$ will be mostly understood from the context onwards. From the collection of RUs in $s(m)$, we obtain a network covering $n(s(m))$ UUs and naturally, any reasonable choice of N must satisfy: $N \geq n(s(m))$.

In order to accompany the computations to follow, we start with an example right from the beginning. For this, we refer to the example of population network in Table 5.1 involving $M = 10$ RUs and resulting in $N = 38$ UUs. Earlier, we discussed about sampling from the population network and took $m = 3$ and ended up with a total of 21 UUs in the sample obtained from the sample RUs, viz. RU_3, RU_5, RU_9. As a continuation to this, let us take 2 more RUs from the remaining 7 RUs in the population by SRSWOR(7, 2). Suppose we obtain RU_6, RU_2. In a sense, we are considering SRSWOR(10, 5) and we end up with RU_2, RU_3, RU_5, RU_6, RU_9. We have already taken care of RU_3, RU_5, RU_9 in Table 5.3. Here we extend that table and incorporate RU_2, RU_6. We find that altogether there are 28 UUs inducted through this sample network based on a sample of 5 RUs (Table 5.5).

At this stage, we also take up appropriate part of Table 5.4 and duly extend it by updating the information on additional induced 7 UUs via RU_2 and RU_6 (Table 5.6).

Our objective now is to study this data set [realized via the network] and come up with a 'reasonable' estimate of N, total number of UUs in the population without any probing. We dispense with unbiasedness and just look for 'reasonably acceptable' estimate for N. Naturally, we demand $N \geq 28$.

Table 5.5 Extended: Sample of UUs induced by selected RUs under SRSWOR($M = 10, m = 5$)

Sampled RUs	Induced UUs
RU_3	[(1, 2, 3, 4, 5, 6, 7)]
RU_5	[(4; 8, 9, 10, 11, 12, 13, 14, 15)]
RU_9	[(1, 2, 3, 7, 11; 16, 17, 18, 19, 20, 21)]
RU_2	[(2, 9); (22, 23, 24)]
RU_6	[(7, 13, 15, 16, 24); (25, 26, 27, 28)]

Table 5.6 Induced UUs and their incidence counts

Induced UUs	Incidence counts among sampled RUs
UUs_1	$r_1 = 2$
UUs_2	$r_2 = 2$
UUs_3	$r_3 = 2$
UUs_4	$r_4 = 2$
UUs_5	$r_5 = 1$
UUs_6	$r_6 = 1$
UUs_7	$r_7 = 2$
UUs_8	$r_8 = 1$
UUs_9	$r_9 = 1$
UUs_{10}	$r_{10} = 1$
UUs_{11}	$r_{11} = 2$
UUs_{12}	$r_{12} = 1$
UUs_{13}	$r_{13} = 1$
UUs_{14}	$r_{14} = 1$
UUs_{15}	$r_{15} = 1$
UUs_{16}	$r_{16} = 1$
UUs_{17}	$r_{17} = 1$
UUs_{18}	$r_{18} = 1$
UUs_{19}	$r_{19} = 1$
UUs_{20}	$r_{20} = 1$
UUs_{21}	$r_{21} = 1$
UUs_{22}	$r_{22} = 1$
UUs_{23}	$r_{23} = 1$
UUs_{24}	$r_{24} = 2$
UUs_{25}	$r_{25} = 1$
UUs_{26}	$r_{26} = 1$
UUs_{27}	$r_{27} = 1$
UUs_{28}	$r_{28} = 1$

Toward this, henceforth, we will regard the above sample network based on 5 RUs as a COMPLETE NETWORK (CN) in its own right based on the 5 RUs sampled in the process. All subsequent studies and related computations are based on this realized CN. We will proceed through the following steps.

Table 5.7 Computations of increments at first level

(Starting RU, Follow-up RU)	Increment
(RU_3, RU_5)	8
(RU_3, RU_9)	7
(RU_3, RU_2)	4
(RU_3, RU_6)	8
**	**
(RU_5, RU_3)	6
(RU_5, RU_9)	10
(RU_5, RU_2)	4
(RU_5, RU_6)	7
**	**
(RU_9, RU_3)	3
(RU_9, RU_5)	8
(RU_9, RU_2)	4
(RU_9, RU_6)	7
**	**
(RU_2, RU_3)	6
(RU_2, RU_5)	8
(RU_2, RU_9)	10
(RU_2, RU_6)	8
**	**
(RU_6, RU_3)	6
(RU_6, RU_5)	7
(RU_6, RU_9)	9
(RU_6, RU_2)	5

Step 1 : Select one RU from the CN. There are 5 choices. For each choice, extend it to one more by choosing a second RU from CN. There are 4 choices. Consequently, there are $5 \times 4 = 20$ ordered pairs formed in the process. For example, we have a pair RU_3 followed by RU_5. Since all ordered pairs [and subsequenty all ordered triplets and higher order -ples] are involved, we will follow the natural order as shown in the table. Anyway, for any such ordered pair, say RU_3, RU_5, we simply count the cardinality of RU_3 [which is the number of induced UUs, i.e., 7] followed by the "additional" number of induced UUs arising out of RU_5, which is 8 in this case. All these 20 ‘increments’ are carefully computed with reference to the CN. These are given in Table 5.7.

Lastly, in *Step* 1, we show the increment frequencies for a total of 20 values as

$$(x,\ f) : [(3, 1), (4, 3), (5, 1), (6, 3), (7, 4), (8, 5), (9, 1), (10, 2)],\ \sum x \times f = 135,\ \text{mean} = 6.75.$$

Step 2 : For the next leg of selection of an RU and computation of the corresponding 'increment' from *Level* 2 to *Level* 3 [i.e., from a 'pair' to a 'triplet'] , note that consideration of a third RU, after having selected two in an ordered fashion, will not depend on the order in which the first two have been chosen. Therefore, there are a total of 10 unordered choices of the pair of RUs. To each such pair, we now associate a third RU out of the remaining three RUs in the CN. Hence we are now in a position to generate 30 triplets and compute 30 increments from *Level* 2 to *Level* 3. One must take due care in doing the necessary computations. These are given in Table 5.8.

Lastly, in *Step* 2, we show the increment frequencies for a total of 30 values as

$$(x,\ f) : [(2, 1), (3, 7), (4, 1), (5, 5), (6, 6), (7, 6), (8, 3), (9, 1)],\ \text{sum}\, x \times f = 163,\ \text{mean} = 5.43.$$

Step 3 : For the next leg of selection of an RU and computation of the corresponding 'increment' from *Level* 3 to *Level* 4, note that consideration of a fourth RU, after having selected three, will not depend on the order in which these have been chosen. Therefore, there are a total of 10 unordered choices of the triplet of RUs. To each such triplet, we now associate a fourth RU out of the remaining two RUs in the CN. Hence we are now in a position to generate 20 increment values from *Level* 3 to *Level* 4. One must take due care in doing the necessary computations. These are given in Table 5.9.

Lastly, in *Step* 3, we show the increment frequencies for a total of 20 values as

$$(x,\ f) : [(2, 4), (3, 3), (4, 1), (5, 6), (6, 5), (7, 1)],\ \text{sum}\, x \times f = 88,\ \text{mean} = 4.4.$$

Step 4 : For the last leg of selection of the RU and computation of the corresponding 'increment' from *Level* 4 to *Level* 5, note that consideration of the fifth RU, after having selected four, will not depend on the order in which these have been chosen. Therefore, there are a total of 5 unordered choices of the 4 RUs. To each such choice, we now associate the last RU in the CN. Hence we are now in a position to generate 5 increment values from *Level* 4 to *Level* 5. One must take due care in doing the necessary computations. These are given in Table 5.10.

Lastly, in *Step* 4, we show the increment frequencies for a total of 5 values as

$$(x,\ f) : [(2, 2), (4, 1), (5, 2)],\ \sum x \times f = 18, \text{mean} = 3.5.$$

In the above, starting with a sample of *RU* of size 4 from the CN, we moved on to sample of size 5 each, augmenting the result with inclusion of the last RU [service center] from within the CN. Then we computed the distribution of the increment in the number of UUs induced through augmentation. This was the last exercise carried out in this context.

Table 5.8 Computations of increments at second level

(Starting RUs Unordered Pair), Follow-up RU	Increment
$(RU_3, RU_5), RU_9$	6
$(RU_3, RU_5), RU_2$	3
$(RU_3, RU_5), RU_6$	6
$(RU_3, RU_9), RU_5$	7
$(RU_3, RU_9), RU_2$	4
$(RU_3, RU_9), RU_6$	7
$(RU_3, RU_2), RU_5$	7
$(RU_3, RU_2), RU_9$	7
$(RU_3, RU_2), RU_6$	7
$(RU_3, RU_6), RU_5$	6
$(RU_3, RU_6), RU_9$	5
$(RU_3, RU_6), RU_2$	3
**	**
$(RU_5, RU_9), RU_3$	2
$(RU_5, RU_9), RU_2$	3
$(RU_5, RU_9), RU_6$	5
$(RU_5, RU_2), RU_3$	5
$(RU_5, RU_2), RU_9$	9
$(RU_5, RU_2), RU_6$	6
$(RU_5, RU_6), RU_3$	5
$(RU_5, RU_6), RU_9$	8
$(RU_5, RU_6), RU_2$	3
**	**
$(RU_9, RU_2), RU_3$	3
$(RU_9, RU_2), RU_5$	7
$(RU_9, RU_2), RU_6$	6
$(RU_9, RU_6), RU_3$	3
$(RU_9, RU_6), RU_5$	8
$(RU_9, RU_6), RU_2$	3
**	**
$(RU_2, RU_6), RU_3$	5
$(RU_2, RU_6), RU_5$	6
$(RU_2, RU_6), RU_9$	8

Table 5.9 Computations of increments at third level

(Starting RUs Unordered Triplet), Follow-up RU	*Increment*
$(RU_3, RU_5, RU_9), RU_2$	3
$(RU_3, RU_5, RU_9), RU_6$	5
**	**
$(RU_3, RU_5, RU_2), RU_9$	6
$(RU_3, RU_5, RU_2), RU_6$	6
**	**
$(RU_3, RU_5, RU_6), RU_9$	5
$(RU_3, RU_5, RU_6), RU_2$	2
**	**
$(RU_3, RU_9, RU_2), RU_5$	6
$(RU_3, RU_9, RU_2), RU_6$	6
**	**
$(RU_3, RU_9, RU_6), RU_5$	5
$(RU_3, RU_9, RU_6), RU_2$	3
**	**
$(RU_3, RU_2, RU_6), RU_5$	5
$(RU_3, RU_2, RU_6), RU_9$	6
**	**
$(RU_5, RU_9, RU_2), RU_3$	2
$(RU_5, RU_9, RU_2), RU_6$	4
**	**
$(RU_5, RU_9, RU_6), RU_3$	2
$(RU_5, RU_9, RU_6), RU_2$	2
**	**
$(RU_5, RU_2, RU_6), RU_3$	5
$(RU_5, RU_2, RU_6), RU_9$	7
**	**
$(RU_9, RU_2, RU_6), RU_3$	3
$(RU_9, RU_2, RU_6), RU_5$	5

Table 5.10 Computations of increments at fourth level

(Starting RUs unordered), last RU	Increment
$(RU_3, RU_5, RU_9, RU_2), RU_6$	4
$(RU_3, RU_5, RU_9, RU_6), RU_2$	2
$(RU_3, RU_5, RU_2, RU_6), RU_9$	5
$(RU_3, RU_9, RU_2, RU_6), RU_5$	5
$(RU_5, RU_9, RU_2, RU_6), RU_3$	2

Table 5.11 Array distribution of increments at different levels

Level (x)	Increment $y \rightarrow$	2	3	4	5	6	7	8	9	10	Total	Mean
1	Freq.	0	1	3	1	3	4	5	1	2	20	6.75
2	Freq.	1	7	1	5	6	6	3	1	3	0	5.43
3	Freq.	4	3	1	6	5	1	0	0	0	20	4.4
4	Freq.	2	0	1	2	0	0	0	0	0	5	3.5

Table 5.12 Fitting of linear regression in array distribution

Level (x)	Array Size (n)	Array Mean $\bar{y}$	Array variance (s_i^2)	Col(4)/Col(2)[s_i^2/n_i]
1	20	6.75	3.7875	0.1893
2	30	5.43	0.5484	0.0183
3	20	4.4	2.54	0.127
4	5	3.5	2.55	0.51

Following Padmawar and Sinha (2016), we may now introduce the concept of array distribution of the increments at each level of the process of induction of the UUs via the RUs. Table 5.11 summarizes our findings.

Padmawar and Sinha (2016) suggested a study of this pattern of array distributions and the underlying regression of y [the increment] on x [the level at which we are studying the increment]. Since the array distributions provide variance estimation in each such array, we may use the concept of heteroscedastic regression as well. In any case, under linear regression, the slope is strictly negative, as is possibly expected. Since 5 RUs have been studied, we have still 5 more left. Our projection should cover $x = 9$ to eventually predict a value of N.

The following computations are now in order (Table 5.12).
Estimating Equations :

$$\bar{y}(x_i) = \alpha + \beta x_i + e_i, \quad i = 1, 2, 3, 4$$

with

$$\hat{V}(\bar{y}(x_i)) = s_i^2/n_i = \text{Col}(4)/\text{Col}(2).$$

Generalized Least Squares Equations:

$$\sum n_i \bar{y}_i/s_i^2 = \alpha \sum n_i/s_i^2 + \beta \sum n_i x_i/s_i^2 \sum n_i \bar{y}_i x_i/s_i^2 = \alpha \sum n_i x_i/s_i^2 + \beta \sum n_i x_i^2/s_i^2.$$

Computations:

$$373.8872 = 69.7622\alpha + 146.0374\beta 760.4881 = 146.0374\alpha + 326.1006\beta.$$

Results: $\hat{\alpha} = 7.6354, \hat{\beta} = -1.0873$.

Prediction: For $x = [5, 6, 7, 8, 9]$, predicted increment-values $= \hat{\alpha} + \hat{\beta} \times [5, 6, 7, 8, 9] = [2.1989, 1.1116, 0.0243, -1.063, -2.1503]$ and this is to be interpreted as $2, 1, 0, 0, 0$.

Summary:

At Level-1, average induced number of UUs = 15 and expected increment is 6 which amounts to 21 at Level-2. Level-2 to Level-3 increment is on an average 4 which pushes the number of UUs to 25. Next, from Level-3 to Level-4, average increment is 4 which pushes the induced number of UUs to 29 at Level-5. Subsequent increments then push this number to $29 + 2 + 1 + 0 + 0 + 0 = 32$ which is our predicted value of N.

Since the variances are not the same across different arrays, we may possibly model the error variances and use suitably defined variance-stabilizing transformations.

Much of the material presented in this chapter is taken from recent research findings, though the subject matter has drawn attention of survey sampling experts/practitioners long time back. Vide Goodman (1953), Darroch (1958), Boswell et al. (1988) and Seber (1982). We have already elaborated contributions of Kiranandan (1976), Maiti et al. (1992) and Sinha et al. (2006) in addressing this problem. It is mentioned that latest thoughts are due to Padmawar and Sinha (2016). Two other papers [Sengupta and De (1997) and De and Sengupta (2003)] are also worth mentioning in this context.

References

Birnbaum, Z. W., & Sirken, M. G. (1965). *Design of sample surveys to estimate the prevalence of rare diseases: Three unbiased estimates*. Vital and health statistics. Series 2, data evaluation and methods research, No. 11 (pp. 1–8).

Boswell, M. T., Burnham, K. P., & Patil, G. P. (1988). Role and use of composite sampling and capture-recapture sampling in echological studies. In P. R. Krishnaiah & C. R. Rao (Eds.), *Handbook of statistics* (Vol. 6, pp. 469–488). Amsterdam: North Holland.

Darroch, J. N. (1958). The multiple recapture census: I Estimation of a closed population. *Biometrika, 45*, 343–359.

De, M., & Sengupta, S. (2003). Bernoulli sequential estimation of the size of a finite population. *Sequential Analysis, 22*, 95–106.

Goodman, L. A. (1953). Sequential sampling tagging for population size problems. *Annals of Mathematical Statistics, 24*, 56–69.

Hedayat, A. S., & Sinha, Bikas K. (1991). *Design and inference in finite population sampling*. NewYork: Wiley.

Kiranandan, S. S. (1976). *Imperfect frames in sample surveys* (Unpublished Ph. D. Thesis, Harvard University).

Maiti, P., Pal, M., & Sinha, B. K. (1992). Estimating unknown dimension(s) of a binary matrix with application to estimation of the size of a mobile population. In S. K. Basu & B. K. Sinha (Eds.), *Probability and Statistics* (pp. 220–233). Narosa Publishing House.

Padmawar, V. R., & Sinha, Bikas K. (2016). Estimating the size of a population comprising of individuals with indirect accessibility. *Journal of Indian Statistical Association, 53*(1–2), 175–200.

Seber, G. A. F. (1982). *The estimation of animal abundance and related parameters* (2nd ed.). New York: Macmillan.

Sengupata, S., & De, M. (1997). On the estimation of size of a finite population. *Sankhya, Series B, 9*, 66–75.

Sinha, Bikas K., Maiti, P., & Sengupta, S. (2006). Unbiased estimation in a finite population: case of multiple indirect identifiability of population units. *Journal of Statistical Theory and Applications, 5*(1), 81–90.

Sirken, M. (1970). Household surveys with multiplicity. *Journal of American Statistical Association, 65*, 257–266.

Chapter 6
Unbiased Estimation of Reliability in Exponential Samples

Learning Objectives
Reliability plays a very important role in judging the performance of a product. Much studies have been conducted to estimate the reliability of a product as well as the stress-strength reliability of a system under different data-generating scenarios, using different estimation methods. In such studies, the use of exponential life time distribution is very common.

The objective of this chapter is to acquaint the readers with the techniques of unbiased estimation of reliability in the exponential set-up. This chapter is rich in contents and provides research-minded readers with enough information in terms of analytical tools and techniques.

6.1 Introduction

In the earlier chapters, we have discussed selection of samples from a finite population of individuals or dwelling units or such. There are, however, situations where sampling is to be performed in an assembly line of an industrial process to determine the reliability of the process. In this chapter, we discuss the problem of estimation of reliability. We concentrate only on the exponential lifetime distribution.

To start with, let us look at the usefulness of reliability in a production process. Reliability plays a very important role in defining the quality of an existing product, and also for the development of new products. For any manufactured product, its reliability is defined as the probability that the product will function satisfactorily for at least a given period of time, under the stated conditions of its usage. There is an underlying assumption in this definition and computation of the reliability at a specified time point. It is tacitly assumed that the 'life' of a product produced [through the production process] can be tested by putting it on test on a time scale till it fails. Naturally the mean life and other features of the 'life distribution' of the

© The Author(s), under exclusive license to Springer Nature Singapore Pte Ltd. 2024
M. Pal and B. K. Sinha, *Selected Topics in Statistical Inference*,
https://doi.org/10.1007/978-981-97-2592-2_6

product play a major role in the evaluation of reliability of the product. This is simply stated, but in reality, it is far from being so simple formulation of the 'experiment' on life distributions. As we proceed, the readers will find considerable variation in this formulation and a host of inferential results, depending on the nature of outcomes.

The lifetime of most products is a random variable. To define it, a commonly used theoretical distribution is the one-parameter exponential distribution. The density function of the distribution is given by

$$f(x) = \frac{1}{\mu} e^{-x/\mu}, \quad x \geq 0, \tag{6.1.1}$$

where μ is the mean lifetime of the product.

Denoting the lifetime by X, we may write $X \sim \text{Exp}(\mu)$.

Under the above lifetime distribution, the reliability of the product at a time point 't' is given by

$$R(t) = \Pr[X > t] = e^{-\frac{t}{\mu}}, \quad t > 0. \tag{6.1.2}$$

When μ is unknown, one would be interested to estimate it, and also estimate $R(t)$, from the lifetimes of a random sample of products put on test. We first focus our attention on unbiased estimation of μ, and then discuss the problem of estimating $R(t)$ unbiasedly.

6.2 Unbiased Estimation of μ

Let n random products, arising out of the same lot and serially numbered as 1, 2, ..., n, be put on life test, and let $X_1, X_2, \ldots, X_n$ denote, respectively, their lifetimes until each one fails. In other words, it is tacitly assumed that the lifetimes observed are the 'full' lifetimes until failure, that is, without any censoring.

Let, $\overline{X} = \frac{1}{n}\sum_{i=1}^{n} X_i$ be the (sample) mean lifetime. Then, from standard results, we have

Theorem 6.2.1 *$\overline{X}$ is the uniformly minimum variance unbiased estimator (UMVUE) of μ.*

[Vide Rao, 1975].

In many practical situations, a complete life-testing experiment may not be feasible or even desirable from the point of view of time, cost, and its destructive nature. In such cases, the experimenter would look at censored data, which allows termination of the experiment before the lifetimes of all the objects put on test are observed. No doubt, this saves both time and money!

Two commonly known 'censoring procedures' are Type I and Type II censoring. Type I censoring is basically 'time censoring', where the lifetimes of the products on test are observed till a pre-specified time point, known as the censoring time. In the general set-up, each product on test may have a different censoring time. In Type II censoring scheme, a random sample of n objects is put on test until k (> 0) objects fail. In this design, the number of failures k, which determines the precision of the study, is pre-fixed. However, the total duration of the test is random.

A mixture of Type I and Type II censoring schemes, known as hybrid censoring, has also been considered by many authors in their studies. In this scheme, a random sample of n objects is put on test. The test is terminated when a pre-fixed number k ($\geq$ 1) of the objects on test fail, or a pre-fixed time point is reached, whichever is earlier. This censoring scheme was first introduced by Epstein (1954) and it has been used in reliability acceptance test in MIL-STD-781-C (1977). Estimation of parameter and also reliability when the life distribution is exponential has been addressed using only the maximum likelihood method and the Bayesian approach. In this book we, however, present discussions on unbiased estimation based on Type I and Type II censored data in the exponential set-up.

6.2.1 *Unbiased Estimation of μ Under Type I Censoring*

Let us consider the censoring time for all the objects put on test be equal, and equal to X^*, say. In this scheme, the number of objects put on test is n. The objects failing in the interval $[0, X^*]$, say r, will be a discrete random variable taking values from 0 to n. Further, for the objects that do not fail in $[0, X^*]$, the only information we have is that their lifetimes are greater than X^*. In case of no failure in $[0, X^*]$, we set $r =$ 0, which means we do not observe any failure times. Thus, in that case, the lifetimes of all the objects on test are greater than X^*.

Note that r is a binomial count variable with parameters (n, q), where $q = 1 - p$ is the probability of 'failure' (meaning failure of an object in $[0, X^*]$), and is given by $q = \Pr[X < X^*] = 1 - \mathrm{e}^{-X^*/\mu}$, and $p = \Pr[X > X^*] = \mathrm{e}^{-X^*/\mu}$ is the probability that an object does not fail in $[0, X^*]$. Then, $E(r) = nq$ and $\mathrm{Var}(r) = npq$.

Let $X_{(0)} = 0$, and $X_{(1)} < X_{(2)} < \cdots < X_{(r)}$ denote the ordered lifetimes of the products failing in $[0, X^*]$. To find the joint probability function $g(r, x_{(0)}, x_{(1)}, \ldots, x_{(r)})$ of $(r, X_{(0)}, X_{(1)}, X_{(2)}, \ldots, X_{(r)})$, we note the following:

(i) For $r = 0$, there are no failures in the interval $[0, X^*]$. Then, the joint probability function of $(r, X_{(0)}, X_{(1)}, X_{(2)}, \ldots, X_{(r)})$ will be the same as the probability that all the objects on test have lifetime greater than X^*. This gives

$$g\big(r, x_{(0)}, x_{(1)}, \ldots, x_{(r)}\big) = p^n, \quad \text{for } r = 0.$$

(ii) For $r \geq 1$ failures in $[0, X^*]$, we find that the ordered lifetimes $X_{(1)}, X_{(2)}, \ldots, X_{(r)}$, conditional on r, can be regarded as the order statistics in a random sample of size

r from a right truncated exponential distribution, truncated at X^*. Thus, given r, the joint density of $X_{(1)}, X_{(2)}, \ldots, X_{(r)}$ is $r! \prod_{i=1}^{r} f\left(x_{(i)}\right)/(1 - R(X^*))^r = \frac{r!}{\mu^r} \frac{e^{-\sum_{i=1}^{r} x_{(i)}/\mu}}{q^r}$, $0 < x_{(1)} < x_{(2)} < \cdots < x_{(r)}$.

[Recall that $R(t)$ is the reliability function defined in (6.1.2).]

Then, as r ~ Binomial(n, q), we have

$$g\left(r, x_{(0)}, x_{(1)}, \ldots, x_{(r)}\right) = r! \binom{n}{r} \frac{p^{n-r}}{\mu^r} e^{-\sum_{i=1}^{r} x_{(i)}/\mu}, \quad 0 < x_{(1)} < x_{(2)} < \cdots < x_{(r)},$$

$$\text{if } 1 \le r \le n.$$

Thus,

$$g\left(r, x_{(0)}, x_{(1)}, \ldots, x_{(r)}\right) = p^n, \quad \text{for } r = 0.$$

$$= r! \binom{n}{r} \frac{p^{n-r}}{\mu^r} e^{-\sum_{i=1}^{r} x_{(i)}/\mu}, \quad 0 < x_{(1)} < x_{(2)} < \cdots < x_{(r)},$$

$$\text{if } 1 \le r \le n.$$

It is readily seen that $(r, Z_r = \sum_{i=1}^{r} X_{(i)})$ is sufficient for μ.

Now, when the lifetime of a product is distributed as Exp(μ), X/μ follows an exponential distribution with mean 1. This simplifies finding an unbiased estimator of μ based on uncensored data, as well as on Type II censored data [to be considered shortly]. But, in the case of Type I censored data, such a transformation leads to the censoring time X^* being transformed to X^*/μ which is not measureable as μ is unknown. Thus intuitively it indicates that *no unbiased estimator of μ exists, under Type 1 censoring*!

For a technical proof of the above claim, see Sengupta et al. (2009).

However, it will be shown in Sect. 6.3 that we can still find an unbiased estimator of $R(t)$ based on Type I censored data.

6.2.2 Unbiased Estimation of μ Under Type II Censoring

In Type II censoring scheme, a random sample of n objects is put on test until k (> 0) objects fail. In this design, the number of failures k, which determines the precision of the study, is pre-fixed. However, the total duration of the test is random.

The data consists of observations on the ordered lifetimes $X_{(1)} < X_{(2)} < \cdots < X_{(k)}$ of the k objects failing, and the lifetimes of the remaining $(n - k)$ objects on test are greater than $X_{(k)}$. Type II censoring is characterized by n, the sample size, and k.

Let, $Z_1 = nX_{(1)}$ and $Z_i = (n-i+1)[X_{(i)} - X_{(i-1)}]$, $i = 2, 3, \ldots, k$. Then, from the property of spacings, we have that Z_i's are independently and identically distributed as Exp(μ). Further, we get $\sum_{i=1}^{k} Z_i = \sum_{i=1}^{k-1} X_{(i)} + (n-k+1)X_{(k)}$, which readily suggests that $\frac{1}{k}\sum_{i=1}^{k} Z_i$ is an unbiased estimator of μ.

From the joint density function of $Z_1, Z_2, \ldots, Z_k$, we obtain $\sum_{i=1}^{k} Z_i$ as a complete sufficient statistic for μ, and, therefore, from Lehmann–Scheffé Theorem, we have the following:

Theorem 6.2.2 $\frac{1}{k}\sum_{i=1}^{k} Z_i = \frac{1}{k}\left[\sum_{i=1}^{k-1} X_{(i)} + (n-k+1)X_{(k)}\right]$ *is UMVUE of* μ.

6.3 Unbiased Estimation of Reliability

Let, $X_1, X_2, \ldots, X_n$ be the lifetimes of n random objects from a lot put on *complete* life test, where the distribution of lifetime is Exp(μ).

In the case of uncensored sampling, a simple unbiased estimator of the reliability $R(t)$ at time point t (based on all observations) is given by

$$\hat{R}(t) = \frac{1}{n}\sum_{i=1}^{n} I(X_i > t),$$

where $I(.)$ denotes an indicator function. The variance of the estimator is

$$\mathrm{var}\left[\hat{R}(t)\right] = \frac{1}{n}R(t)[1 - R(t)].$$

The UMVUE of $R(t)$ is

$$\hat{\hat{R}}(t) = \left(1 - \frac{t}{W}\right)_{+}^{n-1}, \tag{6.3.1}$$

where $W = \sum_{i=1}^{n} X_i$, $d_+ = \max\,[d, 0]$, and its variance comes out to be

$$\mathrm{Var}\left[\hat{\hat{R}}(t)\right] = R(t)\left[G_n\left(t^*\right) - R(t)\right],$$

where $t^* = t/\mu$, and $G_n(t^*) = \frac{1}{\Gamma(n)}\int_0^{\infty} \mathrm{e}^{-u}\frac{u^{2n-2}}{(t^*+u)^{n-1}}\mathrm{d}u$ (vide Johnson et al., 1994, Chap. 19).

Remark 6.3.1 Note that under the exponential distribution exp(μ), $R(t) = \mathrm{e}^{-\frac{t}{\mu}}$, as is indicated in (6.1.1). Therefore, $R^2(t) = R(2t)$, and hence UMVUE of $R^2(t)$ is obtainable from (6.3.1) by replacing 't' by '$2t$'. We may denote it by $\hat{\hat{R}}(2t)$. Further, since $\mathrm{Var}[\hat{\hat{R}}(t)] = E\left[\left\{\hat{\hat{R}}(t)\right\}^2\right] - R^2(t) = E\left[\left\{\hat{\hat{R}}(t)\right\}^2\right] - R(2t)$, again from (6.3.1), we obtain:

$$
\begin{aligned}
\widehat{\mathrm{Var}}\left[\hat{R}(t)\right] = \left[\hat{R}(t)\right]^2 - \hat{R}(2t) &= \left(1 - \frac{t}{W}\right)_+^{2(n-1)} - \left(1 - \frac{2t}{W}\right)_+^{n-1} \\
&= \left(1 - \frac{t}{W}\right)^{2(n-1)} - \left(1 - \frac{2t}{W}\right)^{n-1}, \quad \text{if } 0 < 2t < W \\
&= \left(1 - \frac{t}{W}\right)^{2(n-1)}, \quad \text{if } 0 < t < W < 2t \\
&= 0, \quad \text{if } 0 < W < t.
\end{aligned}
$$

Let us now consider a Type II censored sample, with parameters (n, k) from Exp(μ). Recall that, $Z_1 = nX_{(1)}$, and $Z_i = (n - i + 1)\sum_{i=1}^{k}[X_{(i)} - X_{(i-1)}]$, $i = 2, 3, \ldots, k$, are independently and identically distributed as exponential variables with the same mean μ. So, the UMVUE of the reliability function $R(t)$ is (vide Johnson et al., 1994, Chap. 19)

$$
\hat{R}_1(t) = \left(1 - \frac{t}{W_k}\right)_+^{k-1}, \tag{6.3.2}
$$

where $W_k = \sum_{i=1}^{k} Z_i = \sum_{i=1}^{k-1} X_{(i)} + (n - k + 1)X_{(k)}$.

The variance of the estimator is $\mathrm{var}[\hat{R}_1(t)] = R(t)[G_k(t^*) - R(t)]$, with $t^* = t/\mu$, and $G_k(t^*) = \frac{1}{\Gamma(k)}\int_0^{\infty} \mathrm{e}^{-u}\frac{u^{2k-2}}{(t^*+u)^{k-1}}\mathrm{d}u$.

The above results are readily derivable for a set of observations from a sample of size n. The main challenge, however, lies in finding an unbiased estimator, and thereby the UMVUE, of the reliability function based on Type I censored data. We may base our estimator only on r, the number of failures observed, or on both r and the observed lifetimes.

Let X^* be the censoring time, and let us write t/X^* as s. Then, recall that the reliability at t is given by $R(t) = \mathrm{e}^{-t/\mu}$, which can be written as $R(t) = q^s$, where $q = \mathrm{e}^{-X^*/\mu}$.

6.3.1 Unbiased Estimator of R(t) Based only on r

Suppose that the time points at which failures occurred in the interval $[0, X^*]$ are not recorded, or are ignored. This may happen if the experiment is not continuously monitored, but only the number of failures is recorded at the end of the censoring time. In this case, we look for an unbiased estimator of $R(t)$ based only on r. Recall that $s = t/X^*$.

Case I Suppose s is a positive integer. In this case, the interval $[0, t]$ can be divided into s intervals, each of length X^* as follows:

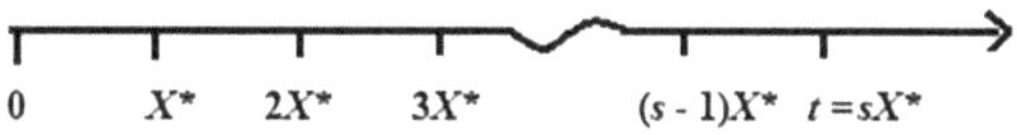

By standard results, it follows that $R(t)$ can be estimated unbiasedly based on r provided $s \leq n$. To obtain an unbiased estimator, we note that for a random variable $X \sim$ Binomial (n, p), $E[X(X-1)\cdots(X-k+1)] = n_{(k)}p^k$, where $1 \leq k \leq n$, k being an integer. In the present case, $(n - r) \sim$ Binomial (n, q), where, as we recall, $q = \mathrm{e}^{-X^*/\mu}$. Thus, we get an unbiased estimator of $R(t) = q^s$ as

$$\hat{R}(t) = 1, \qquad \text{if } r = 0$$
$$= \frac{(n-r)_{(s)}}{n_{(s)}}, \quad \text{if } 1 \leq r \leq n, \tag{6.3.3}$$

where $x_{(c)} = x(x-1)(x-2)\ldots(x-c+1)$, c being a positive integer.

Incidentally, this is the unique unbiased estimator, and hence the UMVUE, of $R(t)$.

For further discussions on the UMVUE, see Remark 6.3.2 below.

Case II For $s = t/X^*$ $(\leq n)$ not an integer, we can still find an unbiased estimator of $R(t)$.

Let s_0 be the smallest integer greater than or equal to s, and r_0 be the number of failures observed up to the time point $T_0 = t/s_0$. Clearly, $T_0 \leq X^*$. In this case, the interval $[0, t]$ is divided in the following way:

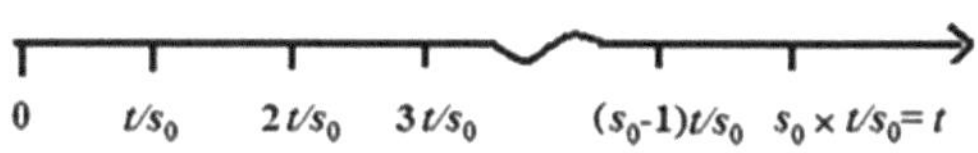

Then arguing as in Case I we get an unbiased estimator of $R(t)$ as

$$\hat{\hat{R}}(t) = 1, \qquad \text{if } r_0 = 0$$
$$= \frac{(n-k_0)_{(r_0)}}{n_{(r_0)}}, \quad \text{if } 1 \leq r_0 \leq n. \tag{6.3.4}$$

Remark 6.3.2 Note that in both the cases above, we have used known results in the context of estimation of the proportion 'q' and its positive integral power(s) [in the set-up of a binomial distribution]. However, the appropriate framework for use of binomial distribution had to be tacitly developed—starting from the framework of a life distribution and the truncation time. Incidentally, we are able to handle both the Cases I and II under consideration by suggesting appropriate context(s) for use of binomial distribution(s). For Case I, we assume 'r' as our available number of failures, and for Case II, we tacitly assume that the number of failures can be observed up to the time point $T_0 = t/s_0$ $(< X^*)$.

6.3.2 Unbiased Estimator of R(t) based on both r and the Observed Lifetimes

Suppose the lifetimes of the r items that fail up to the censoring point X^* are available. Then, as indicated earlier, $(r, Z_r = \sum_{i=1}^{r} X_{(i)})$ is sufficient for μ.

For $n \geq 2$, the unbiased estimator in (6.3.4) is not a function of the sufficient statistic $(r, Z_r = \sum_{i=1}^{r} T_{(i)})$, and hence can be improved upon by Rao–Blackwellization. This and other relatively difficult results are presented in Sengupta et al. (2009). We skip the presentations. Of particular interest to us is Theorem 6.3.1 stated below.

Theorem 6.3.1 *For exponential lifetime distribution, $R(t)$ is unbiasedly estimable based on Type I censored sample if and only if $t \leq nX^*$.*

Remark 6.3.3 Consider the case $X^* = 10$ hours and $n = 10$. It is readily seen that the usual data range is between 0 and 100 h for unbiased estimation of $R(t)$, for any t between 0 and 100 hours. Beyond 100 hours, whether $t = 120$ hours or 125 hours, it really does not matter. In such a situation, no unbiased estimator will exist for $R(t)$.

As indicated earlier, it is difficult to get an explicit expression of $\hat{\hat{R}}^*(t) = E[\hat{\hat{R}}(t)|r, Z_r]$, where $\hat{\hat{R}}(t)$ is the unbiased estimator of $R(t)$ given by (6.3.4). So, one should look for an alternative unbiased estimator based on the sufficient statistic. Necessary and sufficient conditions for the existence of a class of unbiased estimators $g(r, Z_r)$ of $R(t)$ based on the sufficient statistic (r, Z_r), for $0 \leq r \leq n, t \leq nX^*$, are available. Interested readers are referred to Sengupta et al. (2009) for the same.

6.4 Unbiased Estimation of Reliability Based on a Few Selected Order Statistics

In many practical situations, it is quite possible that only k $(1 \leq k \leq n)$ order statistics, say $X_{i_1} < X_{i_2} < \cdots < X_{i_k}$, $1 \leq i_1 < i_2 < \ldots < i_k \leq n$, are available out of the complete set of ordered lifetimes $X_{(1)} < X_{(2)} < \cdots < X_{(n)}$ of n products put on test. This could happen if the experimenter is not regular in observing the failure times. Some particular cases of interest would be as given below:

(1) Initial k order statistics, namely $X_{(1)} < X_{(2)} < \cdots < X_{(k)}$ are available. This is the case of Type II censoring where we stop after the kth failure;
(2) Only one, say the ith order statistic $X_{(i)}$, $i = 2, 3, \ldots, n$, is available;
(3) k consecutive order statistics, starting from the ith one, namely $X_{(i)} < X_{(i+1)} < \cdots < X_{(i+k-1)}$, $2 \leq i \leq n - k + 1$, are available;
(4) k non-consecutive order statistics, namely $X_{(i_1)} < X_{(i_2)} < \cdots < X_{(i_k)}$, where $i_{s+1} \neq i_s + 1$, $s = 1, \ldots, k - 1$, are available.

Finding unbiased estimator of reliability based on Type II censored data has already been discussed. We shall address the problem of unbiased estimation of $R(t)$ in situations 2 and 3. We do not discuss the fourth situation but refer the readers to Sinha et al. (2006) for a detailed deliberation on it.

We consider the lifetime distribution to be Exp(μ).

6.4.1 *Unbiased Estimation of R(t) Based on a Single Order Statistic*

Suppose only the ith order statistic $X_{(i)}$, $1 \leq i \leq n$, is observed. Let us first consider $i = 1, 2$.

For $i = 1$, the unbiased estimator based on the first-order statistic $X_{(1)}$, has the form $I(Z_{(1)} > t)$, where $Z_{(1)} = nX_{(1)}$, and its variance is given by $R(t)[1 - R(t)]$. This result is easy to derive.

For $i = 2$, the density function of $X_{(2)}$ is given by

$$f_{X_{(2)}}(x) = n(n-1)\left[1 - \mathrm{e}^{-x/\mu}\right]\mathrm{e}^{-(n-1)x/\mu}/\mu, \quad 0 < x < \infty.$$

Let us set $Z_2 = (n-1)X_{(2)}$. Then, after routine computation, we obtain

$$\Pr[Z_2 > t] = n\left[\overline{F}(t) - \overline{F}(\alpha t)/\alpha\right], \quad 0 < t < \infty,$$

where $F(.) = 1 - \mathrm{e}^{-x/\mu}$ is the cdf of the parent distribution, $\overline{F}(t) = 1 - F(t)$ and $\alpha = \frac{n}{n-1}$. This gives

$$\overline{F}(t) = \frac{1}{n}\Pr[Z_2 > t] + \overline{F}(t)(\alpha t)/\alpha. \tag{6.4.1}$$

Likewise,

$$\overline{F}(\alpha t)/\alpha = \frac{1}{n\alpha}\Pr[Z_2 > \alpha t] + \overline{F}\left(\alpha^2 t\right)/\alpha^2$$
$$\overline{F}\left(\alpha^2 t\right)/\alpha^2 = \frac{1}{n\alpha^2}\Pr\left[Z_2 > \alpha^2 t\right] + \overline{F}\left(\alpha^3 t\right)/\alpha^3$$

and so on.

Substituting all these in the expression in (6.4.1), we derive

$$\overline{F}(t) = \frac{1}{n}\Pr[Z_2 > t] + \frac{1}{n\alpha}\Pr[Z_2 > \alpha t] + \frac{1}{n\alpha^2}\Pr\left[Z_2 > \alpha^2 t\right] + \cdots.$$

This suggests

$$\hat{\hat{F}}(t) = \frac{1}{n}I[Z_2 > t] + \frac{1}{n\alpha}I[Z_2 > \alpha t] + \frac{1}{n\alpha^2}I\left[Z_2 > \alpha^2 t\right] + \cdots$$
$$= \sum_{j=0}^{\infty} \frac{1}{n\alpha^j} I\left(Z_2 > \alpha^j t\right). \tag{6.4.2}$$

Since Z_2-distribution belongs to the *complete class* of exponential families, it follows that (6.4.2) provides the *unique unbiased estimator* based on the second-order statistic $X_{(2)}$, for $R(t)$.

To work out the variance of $g_2(X_{(2)})$, we go by the definition

$$\text{Var}\left[g_2\left(X_{(2)}\right)\right] = E\left[\left\{g_2\left(X_{(2)}\right)\right\}^2\right] - [R(t)]^2.$$

Now, $\{g_2(X_{(2)})\}^2$ involves squares and cross products of indicator functions. Since $\alpha > 1$, we get

$$I\left[Z_2 > \alpha^j t\right] \times I\left[Z_2 > \alpha^k t\right] = I\left[Z_2 > \alpha^k t\right], \quad \text{whenever } k > j \geq 0.$$

So, all square and cross product terms are easily expressible in terms of indicator functions before taking expectations. Routine simplification then leads to

$$\text{Var}\left[g_2\left(X_{(2)}\right)\right] = \frac{1}{n}\left[R(t) + \frac{1}{\alpha}\left(1 + \frac{1}{\alpha}\right)R(\alpha t) + \frac{1}{\alpha^3}\left(1 + \frac{1}{\alpha}\right)R\left(\alpha^2 t\right) + \cdots\right] - [R(t)]^2.$$

Some interesting properties of the estimator $g_2(X_{(2)})$ are given below without proof:

(i) For $n \geq 2$, $\text{Var}[g_2(X_{(2)})] < \text{Var}[g_1(X_{(1)})]$, uniformly in α.
(ii) For $n = 2$, $\text{Var}[g_2(X_{(2)})] < \frac{1}{2}\text{Var}[g_1(X_{(1)})]$, uniformly in α.
(iii) An unbiased estimator of $\text{Var}[g_2(X_{(2)})]$ is given by

$$\hat{V}\left[g_2\left(X_{(2)}\right)\right] = g_2^2\left(X_{(2)}\right) - g_2\left(X_{(2)}/2\right),$$

which can be shown to be uniformly non-negative for $n = 2$.

Remark 6.4.1 The case of $R(t)$ estimation based on $X_{(2)}$ has been presented above. The readers will find the derivation interesting in view of the representation (6.4.1) made. This leads to (6.4.2) by successive use. For higher ordered single statistic, we go by a different technique which is elaborated below.

Let $g_i(X_{(i)})$ be an unbiased estimator of $R(t)$ based on $X_{(i)}$ $(1 \leq i \leq n)$. Then, by definition, for all $0 < \mu < \infty$,

$$
\begin{aligned}
\frac{1}{\mu}\int_0^\infty I(x>t)\mathrm{e}^{-x/\mu}\mathrm{d}x &= R(t) = E_\mu\left[g_i\left(X_{(i)}\right)\right] \\
&= \frac{1}{\mu}\int_0^\infty g_i(x)\frac{\mathrm{e}^{-(n-i+1)/\mu}}{B(i,n-i+1)}\left(1-\mathrm{e}^{-x/\mu}\right)^{i-1}\mathrm{d}x \\
&= \frac{1}{\mu}\int_0^\infty \sum_{j=0}^{i-1}(-1)^j\binom{i-1}{j}g_i(x)\frac{\mathrm{e}^{-(n-i+j+1)/\mu}}{B(i,n-i+1)}\mathrm{d}x \\
&= \frac{1}{\mu}\int_0^\infty \sum_{j=0}^{i-1}(-1)^j\binom{i-1}{j}g_i\left(\frac{x}{n-i+j+1}\right) \\
&\quad \frac{\mathrm{e}^{-x/\mu}}{(n-i+j+1)B(i,n-i+1)}\mathrm{d}x,
\end{aligned}
$$

where $B(\alpha, \beta)$ is the beta function.

Comparing the two sides, we therefore have that the unbiased estimator satisfies the identity

$$
I(x>t) = \frac{1}{B(i,n-i+1)}\sum_{j=0}^{i-1}(-1)^j\binom{i-1}{j}\frac{g_i\left(\frac{x}{n-i+j+1}\right)}{n-i+j+1}. \tag{6.4.3}
$$

Further, from the completeness of the exponential distribution, it follows that the estimator is unique.

Expression (6.4.3) is fundamental to the study of UMVUE of $R(t)$ based on a single order statistic. One can deduce the expression (6.4.2) from here. It is instructional to deduce explicit form of the estimator when using $X_{(3)}$ for $n \geq 3$. For details, we refer the readers to Sinha et al. (2006).

6.4.2 Unbiased Estimation of R(t) Based on a Few Order Statistics

Suppose the experimenter has data only on k $(2 \leq k \leq n)$ consecutive order statistics, starting from the ith one, viz. $X_{(i)}, X_{(i+1)}, \ldots, X_{(i+k-1)}$, $i, k \geq 2$, $i+k \leq n+1$, obtained from the lifetimes of n random objects put on life test.

To obtain an unbiased estimator of $R(t)$, let us define

$$
Z_j = (n-i-j+1)\left[X_{(j+i)} - X_{(j+i-1)}\right],\quad j = 1, 2, \ldots, k-1.
$$

We note that $Z_1, Z_2, \ldots, Z_{k-1}$ are independently and identically distributed as one-parameter exponential with mean μ and are also distributed independently of $X_{(i)}$. Then, the UMVUE, say $\hat{R}_k(t)$, of $R(t)$ based on the Z_j's is similar to that obtained under Type II censoring, and given by (3.2), except that k is replaced by $k-1$. That is,

$$\hat{R}_k(t) = \left(1 - \frac{t}{W_k^*}\right)_+^{k-2},$$

where $W_k^* = \sum_{i=1}^{k-1} Z_i$, and its variance is given by

$$\operatorname{Var}\big(\hat{R}_k(t)\big) = R(t)\big[I\big(t^*|k\big) - R(t)\big],$$

where $t^* = t/\mu$, and $I(t^*|k) = \frac{1}{\Gamma(k-1)} \int_0^\infty \mathrm{e}^{-u} \frac{u^{2k-4}}{(t^*+u)^{k-2}} \mathrm{d}u$.

Remark 6.4.2 Since $\hat{R}_k(t)$ and $g_i(X_{(i)})$, given by (6.4.3), are independent unbiased estimators of $R(t)$, it is tempting to combine them in some suitable way. Consider $\tilde{R}_{ik}(t) = \frac{1}{2}[g_i(T_{(i)}) + \hat{R}_k(t)]$. This will also be an unbiased estimator of $R(t)$, and its variance is given $\frac{1}{4}[\operatorname{Var}(g_i(T_{(i)})) + \operatorname{var}(\hat{R}_k(t))]$.

6.5 Interval Estimation of Reliability Function

Interval estimation of reliability function is very important to manufacturers as it gives an idea of the extent (in terms of percentages) to which a product may be reliable. In our discussion on interval estimation, we consider the sample size to be large, and obtain bounds on the reliability function based on (a) an unbiased estimator of reliability and (b) the likelihood function. In the latter case, we obtain what may be called likelihood ratio bounds. Interested readers are referred to Chap. 5 in Strimmer (2023).

Suppose the lifetime X of a product has an exponential distribution with mean lifetime $\lambda > 0$ (unknown). Let $X_1, X_2, \ldots, X_n$ be the lifetimes of n products put simultaneously on life test.

For a given $t > 0$, the reliability of the product is

$$R(t) = \Pr \cdot [X > t] = \mathrm{e}^{-t/\lambda}.$$

(a) An unbiased estimate of $R(t)$ is given by

$$\tilde{R}(t) = \frac{1}{n} \sum_{i=1}^{n} I[(X_i > t)],$$

where I is the indicator function.

It is readily seen that

(i) $E(\tilde{R}(t)) = R(t)$.
(ii) $V(\tilde{R}(t)) = \frac{[R(t)-R(2t)]}{n}$, since $R^2(t) = R(2t)$.
(iii) $\tilde{V}(\tilde{R}(t)) = \frac{[\tilde{R}(t)-\tilde{R}(2t)]}{n} = \frac{\tilde{R}(t)[1-\tilde{R}(t)]}{n}$.

For large sample size, we have

$$Z = \frac{\tilde{R}(t) - R(t)}{\sqrt{\tilde{R}(t)(1-\tilde{R}(t))/n}} \text{ is approximately distributed as } N\ (0,\ 1).$$

So, starting with $\Pr\left[|Z| \le Z_{\frac{\alpha}{2}}\right] = 1-\alpha$, we get the confidence interval for $R(t)$ as

$$\tilde{R}(t) \pm Z_{\frac{\alpha}{2}}\sqrt{\tilde{R}(t)(1-\tilde{R}(t))/n}. \tag{6.5.1}$$

For observed value of $\tilde{R}(t)$, we may be interested in knowing how likely is $R(t)$ to differ from its estimate by at most p (a given small fraction). This can be easily answered from the following:

$$\Pr\left[\left|\tilde{R}(t) - R(t)\right| \le p\right] = \Phi\left(\frac{p}{\sqrt{\tilde{R}(t)(1-\tilde{R}(t))/n}}\right) - \Phi\left(\frac{-p}{\sqrt{\tilde{R}(t)(1-\tilde{R}(t))/n}}\right).$$

Table 6.1 gives the 95% confidence interval for $R(t)$, and the chance that $\left|\tilde{R}(t) - R(t)\right| \le 0.05, 0.1$, when $\tilde{R}(t) = 0.6$.

Remark 6.5.1 It must be noted that the above approach does not utilize the nature of distribution of the lifetimes of the items. It is nonparametric in nature. It may be worth deriving an estimate of λ from the data [which is the sample mean in this case]

Table 6.1 Showing the 95% confidence interval for $R(t)$, and the probability that $\left|\tilde{R}(t) - R(t)\right| \le 0.05, 0.1$, when $\tilde{R}(t) = 0.6$

N	Lower confidence limit	Upper confidence limit	Chance that $\left\|\tilde{R}(t) - R(t)\right\| \le 0.05$	Chance that $\left\|\tilde{R}(t) - R(t)\right\| \le 0.10$
50	0.464207	0.735793	0.5295	0.8511
60	0.476039	0.723961	0.5708	0.8862
70	0.485234	0.714766	0.6068	0.9123
80	0.492646	0.707354	0.6387	0.9321
90	0.498786	0.701214	0.6671	0.9472
100	0.50398	0.69602	0.6926	0.9588

and utilizing the functional form of $R(t)$ to deduce its confidence bounds. This is studied below.

(b) The likelihood ratio (LR) confidence bounds for $R(t)$ is obtained in the following way: Based on the random sample $X_1, X_2, \ldots, X_n$ indicated earlier, we can get the maximum likelihood (ML) estimator of λ as $\widehat{\lambda} = \frac{1}{n}\sum_{i=1}^{n} X_i$. Further, we have

$$R(t) = \mathrm{e}^{-t/\lambda},$$

which gives

$$\lambda = -\frac{t}{\ln R(t)}. \tag{6.5.2}$$

The likelihood ratio confidence bounds on $R(t)$ with confidence coefficient $(1 - \alpha)$ are then obtained from the inequality

$$-2\left[\ln L(\lambda) - \ln L\left(\widehat{\lambda}\right)\right] \le \chi^2_{1,\alpha}, \tag{6.5.3}$$

where L denotes the likelihood function, and we replace λ by its functional form in $R(t)$, indicated in (6.5.2).

Note that the degrees of freedom of the Chi-square is equal to the number of parameters estimated, which is 1 in this case.

For equality in (6.5.3) we get the equation

$$\frac{U\sum_{i=1}^{n} X_i}{t} - 2\ln U = \chi^2_{1,\alpha} - 2n\ln t + 2n\ln\widehat{\lambda} + 2n, \tag{6.5.4}$$

where $U = -\ln R(t)$. So, $U \in [0, \infty)$.

It is easy to note that the LHS of (6.5.4) is a strictly convex function of U, while the RHS of (6.5.4) is a positive constant for given t, α and the observations. We, therefore, get two solutions to U satisfying (6.5.4), say, $U_1 < U_2$. Thus, if $0 < U_1 < U_2 < \infty$, the confidence interval for U is $[U_1, U_2]$, and, accordingly, we get the confidence interval for $R(t)$ as $[\mathrm{e}^{-U_2}, \mathrm{e}^{-U_1}]$.

Example Suppose 25 items were put on a life test and their failure times were observed as 10, 11.5, 12, 12, 15, 18, 18.75, 20, 21, 25, 30, 32, 32.7, 34, 34, 37, 39.2, 40, 42, 43.5, 45, 47, 48, 50, and 53.45 hours. Assuming exponential failure time distribution with mean λ (unknown), the ML estimate of λ is obtained as the sample mean $\widehat{\lambda} = 30.844$.

Table 6.2 shows the 95% LR bounds for $R(t)$ along with the 95% confidence bounds for $R(t)$ under asymptotic normality of $\tilde{R}(t)$ for five values of t.

Remark 6.5.2 It may be noted that while $\tilde{R}(t)$ depends on the chosen value of 't',

Table 6.2 Showing the 95% confidence bounds for $R(t)$ using the likelihood ratio method and the method based on $\tilde{R}(t)$

t	$\tilde{R}(t)$	Likelihood ratio bounds		Bounds using $\tilde{R}(t)$	
		Lower bound	Upper bound	Lower bound	Upper bound
10	0.96	0.6260	0.8080	0.8832	1.0368[a]
15	0.80	0.4953	0.7264	0.6432	0.9568
20	0.68	0.3918	0.6529	0.4971	0.8628
35	0.40	0.1940	0.4743	0.2080	0.5920
45	0.16	0.1215	0.3832	0.0163	0.3037

[a] Since $R(t)$ cannot exceed 1, the upper bound is taken to be 1

$\hat{\lambda}$ is just the sample mean, and it is independent of the choice of 't'. Parametric formulation and large sample approximation produced *closer* confidence intervals for $R(t)$ for all tabulated values of 't', except '$t = 10$'.

6.6 Unbiased Estimation of Stress-Strength Reliability Under One-Parameter Exponential Distribution

So far, we have discussed estimation of reliability of a product, which is given by the probability that the product will function for at least a specified period of time under the stated conditions of usage. The distribution of the lifetime is assumed to be one-parameter exponential with unknown mean μ. A product may, however, be subjected to stresses that can hamper its performance, as well as its longevity. A stress is basically a load that tends to lead to the failure of the product. Some common environmental stresses are the weather stresses, like pressure, temperature, humidity, and the location stresses like altitude, power quality, and radiation. In engineering, stress is defined as the pressure developed on application of mechanical loads on a product. To combat the stresses, the product or system should be strong enough to perform satisfactorily in the face of the stresses. Thus, the *strength of a product* is defined as its ability to withstand a stress on it without failure. In most cases, both stress and strength are assumed to be random variables. The variation in the strength may be attributed to random variation in the material properties, in the production process, etc.

Consider a product or system with strength X, which is subjected to a stress Y. The stress-strength reliability of the product is defined as the probability that the strength of the system exceeds the stress on it, that is, $\Pr[X > Y]$.

Probability of the type $\Pr[X > Y]$ may arise in many situations other than stress-strength reliability, like in biometry, medicine, engineering, etc. For example, suppose there are two similar drugs, D_1 and D_2, that are being administered to patients suffering from a certain terminal disease. If X and Y denote, respectively, the lengths of the remaining life of a patient on being administered drugs D_1 and D_2, then, given a choice, a patient would make his choice of drug based on the probability that X exceeds Y. If this probability is greater than 1/2, he would choose drug D_1; if it is less than 1/2, he would choose drug D_2; and if it is equal to 1/2 he would go for any one of them. Again, if X denotes the diameter of a shaft and Y the diameter of a bearing that is to be mounted on the shaft, then $\Pr(X > Y)$ would indicate the chance that the bearing will fit the shaft without any interference.

When strength X and stress Y are distributed independently as $\text{Exp}(\mu)$ and $\text{Exp}(\lambda)$, respectively, the expression for the stress-strength reliability is given by

$$R = \Pr[X > Y] = \int_0^\infty \Pr[X > y]\frac{1}{\lambda}e^{-y/\lambda}dy$$

$$= \int_0^\infty e^{-y/\mu} \cdot \frac{1}{\lambda}e^{-y/\lambda}dy = \frac{\mu}{\mu+\lambda}. \tag{6.6.1}$$

For unknown μ and/or λ, one looks into the problem of estimating the stress-strength reliability. We shall discuss unbiased estimation of stress-strength reliability based on independent random samples $(X_1, X_2, \ldots, X_m)$ and $(Y_1, Y_2, \ldots, Y_n)$ on strength and stress, respectively. On identifying the complete sufficient statistics for the unknown parameters, the UMVUE of R can be obtained using the Lehmann–Scheffé Theorem. We consider all possible cases. It may be noted that Tong (1974, 1975) was the first to consider the problem of finding the UMVUE of R.

Case 1 μ unknown, λ known

In this case, an unbiased estimator of R is given by $U_1 = 1 - e^{-X_1/\lambda}$. It is readily seen that $T_1 = \sum_{i=1}^m X_i$ is a complete sufficient statistic for μ. Hence, the UMVUE of R is given by $E(U_1 \mid T_1)$.

Noting that the density of X_1 given $T_1 = t_1$ is

$$f_1(x_1|t_1) = \frac{m-1}{t_1}\left(1 - \frac{x_1}{t_1}\right)^{m-2}, \quad 0 < x_1 < t_1,$$

the UMVUE of R is given by the theorem below.

Theorem 6.6.1 *If the strength X and stress Y be two independent random variables following the one-parameter exponential distributions Exp(μ) and Exp(λ), respectively, where only μ is unknown, then the UMVUE of R = P(Y > X) is given by*

$$\hat{R}_1 = E(U_1|T_1) = 1 - \sum_{s=0}^{m-2}(-1)^s\frac{(m-1)!}{(m-s-2)!}\left(\frac{\lambda}{T_1}\right)^{s+1}\text{IG}_{\frac{T_1}{\lambda}}(s+1),$$

where $\text{IG}_w(\alpha) = \frac{1}{\Gamma(\alpha)}\int_0^w e^{-z}z^{\alpha-1}dw.$

Proof We have

$$E(U_1|T_1=t_1)=\int_0^{t_1}\left(1-e^{-x_1/\lambda}\right)\frac{m-1}{t_1}\left(1-\frac{x_1}{t_1}\right)^{m-2}dx_1$$

$$=1-(m-1)\int_0^{t_1}e^{-x_1/\lambda}\sum_{s=0}^{m-2}(-1)^s\binom{m-2}{s}\frac{x_1^s}{t_1^{s+1}}dx_1$$

$$=1-\sum_{s=0}^{m-2}(-1)^s\frac{(m-1)!}{(m-s-2)!}\left(\frac{\lambda}{t_1}\right)^{s+1}\frac{1}{\Gamma(s+1)}\int_0^{t_1/\lambda}e^{-z}z^s dz$$

$$=1-\sum_{s=0}^{m-2}(-1)^s\frac{(m-1)!}{(m-s-2)!}\left(\frac{\lambda}{t_1}\right)^{s+1}\mathrm{IG}_{\frac{t_1}{\lambda}}(s+1)\left(\frac{\lambda}{t_1}\right)^{s+1},$$

where $\mathrm{IG}_w(\alpha)$ is the incomplete Gamma function given by $\mathrm{IG}_w(\alpha)=\frac{1}{\Gamma(\alpha)}\int_0^w e^{-z}z^{\alpha-1}dw$.

Case 2 μ known, λ unknown

Here $T_2=\sum_{j=1}^m Y_j$ is a complete sufficient statistic for λ, and an unbiased estimator of R is given by $U_2=e^{-Y_1/\mu}$. Noting that the density function of U_2 given $T_2=t_2$ has a similar expression as that of X_1 given $T_1=t_1$, and proceeding as in the proof of Theorem 6.6.1, we obtain the UMVUE of R as

$$\hat{R}_2=E(U_2|T_2)=\sum_{s=0}^{n-2}(-1)^s\frac{(n-1)!}{(n-s-2)!}\mathrm{IG}_{\frac{T_2}{\mu}}(s+1)\left(\frac{\mu}{T_2}\right)^{s+1},$$

where $\mathrm{IG}_w(\alpha)=\frac{1}{\Gamma(\alpha)}\int_0^w e^{-z}z^{\alpha-1}dw$.

Case 3 Both μ and λ unknown

Here, an unbiased estimator of R is given by

$$\begin{aligned}V&=1,\quad \text{if } X_1>Y_1\\&=0,\quad \text{otherwise.}\end{aligned}$$

Further, $T=(T_1,T_2)$ is a complete sufficient for (μ,λ), where T_1 and T_2 are as defined earlier.

The UMVUE of R is therefore given by $E(V\mid T)$, whose explicit form is indicated in Theorem 6.6.2.

Theorem 6.6.2 *If the strength X and stress Y be independent random variables following the one-parameter exponential distributions given by Exp(*μ*) and Exp(*λ*),* *respectively, and both μ and λ be unknown, then the UMVUE of $R = P(Y > X)$ is given by*

$$\hat{R} = E(V|T) = \sum_{s=0}^{m-1} (-1)^s \frac{(m-1)!(n-1)!}{(m-1-s)!(n-1+s)!} \left(\frac{T_2}{T_1}\right)^s, \quad \text{if } T_2 < T_1$$

$$= 1 - \sum_{s=0}^{n-1} (-1)^s \frac{(m-1)!(n-1)!}{(m-1-s)!(n-1+s)!} \left(\frac{T_1}{T_2}\right)^s, \qquad \text{if } T_1 < T_2.$$

Proof Consider any realization $t = (t_1, t_2)$ of $T = (T_1, T_2)$.

First let us consider $t_1 > t_2$.

Then,

$$\begin{aligned}
E(V|T = t) &= \Pr[X_1 > Y_1 | T = t] \\
&= \int_0^{t_2} \left[\int_{y_1}^{t_1} \frac{m-1}{t_1}\left(1 - \frac{x_1}{t_1}\right)^{m-2} \mathrm{d}x_1 \right] \frac{n-1}{t_2}\left(1 - \frac{y_1}{t_2}\right)^{n-2} \mathrm{d}y_1 \\
&= \int_0^{t_2} \left(1 - \frac{y_1}{t_1}\right)^{m-1} \frac{n-1}{t_2}\left(1 - \frac{y_1}{t_2}\right)^{n-2} \mathrm{d}y_1 \\
&= \sum_{s=0}^{m-1} (-1)^s \binom{m-1}{s} (n-1) \int_0^{t_2} \left(\frac{y_1}{t_1}\right)^s \left(1 - \frac{y_1}{t_2}\right)^{n-2} \mathrm{d}\left(\frac{y_1}{t_2}\right) \\
&= \sum_{s=0}^{m-1} (-1)^s \binom{m-1}{s} (n-1) B(s+1, n-1) \left(\frac{t_2}{t_1}\right)^s \\
&= \sum_{s=0}^{m-1} (-1)^s \frac{(m-1)!(n-1)!}{(m-1-s)!(n-1+s)!} \left(\frac{t_2}{t_1}\right)^s,
\end{aligned}$$

where $B(a, b) = \int_0^1 z^{a-1}(1-z)^{b-1}\mathrm{d}z = \frac{\Gamma(a)\Gamma(b)}{\Gamma(a+b)}$.

Now consider $t_1 < t_2$. Since X_1 cannot exceed t_1, for $Y_1 < X_1$ we must have $0 < Y_1 < X_1 < t_1$. Hence,

$$\begin{aligned}
E(V|T = t) &= \Pr[X_1 > Y_1 | T = t] \\
&= \int_0^{t_1} \left[\int_{y_1}^{t_1} \frac{m-1}{t_1}\left(1 - \frac{x_1}{t_1}\right)^{m-2} \mathrm{d}x_1 \right] \frac{n-1}{t_2}\left(1 - \frac{y_1}{t_2}\right)^{n-2} \mathrm{d}y_1.
\end{aligned}$$

Simplifying the above integral, we obtain

$$E(V|T = t) == 1 - \sum_{s=0}^{n-1} (-1)^s \frac{(m-1)!(n-1)!}{(m-1+s)!(n-1-s)!} \left(\frac{t_1}{t_2}\right)^s.$$

Remark 6.6.1 It is interesting to note that when both the parameters μ and λ are unknown, the UMVUE of the stress-strength reliability is a function of the ratio of the sum of sample strengths to the sum of sample stresses, namely $\frac{T_1}{T_2}$. Thus, the observed value of the ratio affects the estimate of R rather than the individual values of T_1 and T_2.

A heuristic justification for the ratio $\frac{T_1}{T_2}$ playing the central role in the expression of the UMVUE can be given as follows:

Note that basically,

$$2T_1/\mu \sim \chi^2_{2m} \text{ and } 2T_2/\lambda \sim \chi^2_{2n}.$$

Further, for $n \geq 2$,

$$E(T_2) = n\lambda \text{ and } E(1/T_2) = 1/[(n-1)\lambda].$$

Since T_1 and T_2 are independent, an unbiased estimator for μ/λ is derived from

$$E[(n-1)T_1/T_2] = (n-1)\mu/[(n-1)\lambda] = \mu/\lambda,$$

and this is the UMVUE of μ/λ. This follows from Lehmann–Scheffé Theorem since (T_1, T_2) is complete sufficient for (μ, λ).

Now starting with any unbiased estimator of $R(t)$ such as $I(X > Y)$, we consider its conditional expectation, given the sufficient statistics (T_1, T_2). This is as good as working out the conditional expectation, given T_1/T_2. That would result in the UMVUE of $R(t)$ being a function of T_1/T_2.

Remark 6.6.2 In the discussion above, we have assumed stress and strength variables to vary in the interval $[0, \infty)$. However, in many real-life situations, the manufactured products are subjected to at least a minimum (greater than zero) built-in stress, and therefore the manufacturers design their products to also have at least a minimum strength (greater than zero). As such, we have that stress and strength are random variables taking values in the intervals, say, $[a, \infty)$ and $[b, \infty)$, respectively, where $a, b > 0$. When the two-parameter exponential distributions appropriately define the random fluctuations in stress and strength, the problem of unbiasedly estimating the stress-strength reliability is rather involved. We do not indulge in discussion of the highly non-trivial solutions to the problem. However, interested readers are referred to the following for the same: Iwase (1987), Ivshin (1996), Pal et al. (2005).

References

Epstein, B. (1954). Truncated life tests in the exponential case. *Annals of Mathematical Statistics, 25*, 555–564.

Ivshin, V. V. (1996). Unbiased estimators of Pr(X < Y) and their variances in the case of uniform and two-parameter exponential distributions. *Journal of Mathematical Science, 81*, 2790–2793.

Iwase, K. (1987). On UMVUE estimators of Pr(Y < X) in the two-parameter exponential case. *Memoirs of the Faculty of Engineering, Hiroshima University, 9*, 21–24.

Johnson, N. L., Kotz, S., & Balakrishnan N. (1994). *Continuous univariate distributions* (Vol. 1, 2nd ed.). Wiley.

MIL-STD-781-C. (1977). *Reliability design qualifications and production acceptance test, exponential distribution.* U.S. Government Printing Office.

Pal, M., Ali, M. M., & Woo, J. (2005). Estimation and testing of P(Y > X) in two-parameter exponential distributions. *Statistics, 39*(5), 415–428.

Rao, C. R. (1973). *Linear statistical inference and its applications.* Wiley.

Sengupta, S., Sinha, B. K., & Bagchi, S. B. (2009). Estimation of mean life and reliability for exponential life distribution using time censored sample data. *Statistics and Probability Letters, 79*(7), 899–906.

Sinha, B. K., Sengupta, S., & Mukhoti, S. (2006). Unbiased estimation of the distribution function of an exponential population using order statistics with application in ranked set sampling. *Communications in Statistics—Theory and Methods, 35*(9), 1655–1670.

Strimmer, K. (2023). *Statistical methods: Likelihood, Bayes and regression.* https://strimmerlab.github.io/publications/lecture-notes/MATH20802/

Tong, H. (1974). A note on the estimation of P[Y < X] in the exponential case. *Technometrics, 16*, 625.

Tong, H. (1975). Letter to the editor. *Technometrics, 17*, 393.

Exercises

Chapter 2

1. Show that in regular binomial sampling with a fixed sample of size 'n', there does not exist any ***proper*** unbiased estimator for '$p\ (1 - p)$' for any value of $n \geq 2$.
2. Generalize the above non-existence result to the function '$p^2(1 - p)$' for infinitely many values of 'n'. You may start with $n = 5$.
3. Can you modify the above to an existence result for the function '$p^3(1 - p)^2$' for a finite set of values of 'n'?

Chapter 3

4. Consider the set-up of stratified random sampling and follow the usual notations and terminology. Vide, for example, Hedayat and Sinha (1991).

   (a) Starting with $L = 6$ strata with assigned values of $N_h's$, $SS_h's$ and $n = 40$, following Tommy Wright (2017)*, derive the sample sizes $n_h's$ under Neyman optimal allocation rule.

   (*Exact optimal sample allocation: More efficient than Neyman—Statistics & Probability Letters, Volume 129, October 2017, Pages 50–57.)

   (b) Next, consider additional resources in terms of additional $k = 5$ units.

   Further, consider all possible allocations of these 5 extra units covering the strata numbers 1, 3, and 5. Work out variance expressions of the estimate of overall population mean under enhanced stratified sampling designs for all possible choices and compare their performances by ranking them, starting from the best.

© The Editor(s) (if applicable) and The Author(s), under exclusive license to Springer Nature Singapore Pte Ltd. 2024

M. Pal and B. K. Sinha, *Selected Topics in Statistical Inference*,
https://doi.org/10.1007/978-981-97-2592-2

(c) Examine if each of the possible alternative stratified sampling designs fares better than the original stratified sampling design in (a).

Chapter 4

5. Provide a proof of Theorem 4.2.5. Examine if the claim can be strengthened by knowing the variance ratio, instead of both.
6. (a) For a sample of size $n = 2$ from a normal population with known mean and unknown variance, state and prove a result for existence of a sufficient statistic for variance parameter.
   (b) Extend the above result to the case of $n = 3$.
7. In the context of linear regression over [− 1, 1], assume $V(Y_x) = -\frac{2\sigma^2}{1+x^2}$, for $-1 \leq x \leq 1$ with σ^2 known.
   Deduce expressions for sufficient statistics and work out results similar to those in Theorems 4.3.1, 4.3.2, and 4.3.3.
8. Provide a proof of Theorem 4.3.5.
9. Provide a proof of Theorem 4.4.1.

Chapter 5

10. Create a population network of $N = 30$ ultimate units [UUs] and then consider a set of $K = 7$ reference units so that the RUs collectively engulf all the UUs. Follow the sampling methodology for a sample of $k = 3$ RUs and build up the sample network data involving selected RUs and induced UUs.
11. With reference to the context given in Q10, based on the sample network for $k = 3$ RUs as stipulated above, provide an unbiased estimate of 'N', pretending it to be unknown. Study its non-negativity aspect for all possible sample compositions.

Chapter 6

12. Based on a Type II censored data set from a one-parameter exponential lifetime distribution with unknown mean θ, find the maximum likelihood estimator of θ, and use it to find the UMVUE of the reliability function $R(t)$ at a specified time point $t > 0$.
13. Refer to Sect. 6.4.1.
    We have introduced $g_2(X_{(2)})$, and Var $[g_2(X_{(2)}),]$ in this subsection. In this context, establish the claims in (i)–(iii).
14. Examine how the confidence bounds for R based on maximum likelihood estimator compare with the bounds based on asymptotic normality of unbiased estimator of R.
15. Obtain alternative expressions of UMVUE of stress-strength reliability in cases 1–3 under one-parameter exponential distributions of stress and strength.

Author Index

© The Editor(s) (if applicable) and The Author(s), under exclusive license to Springer Nature Singapore Pte Ltd. 2024
M. Pal and B. K. Sinha, *Selected Topics in Statistical Inference*,
https://doi.org/10.1007/978-981-97-2592-2

Subject Index

© The Editor(s) (if applicable) and The Author(s), under exclusive license to Springer Nature Singapore Pte Ltd. 2024

M. Pal and B. K. Sinha, *Selected Topics in Statistical Inference*,
https://doi.org/10.1007/978-981-97-2592-2

MIX
Papier aus verantwortungsvollen Quellen
Paper from responsible sources
FSC® C105338

If you have any concerns about our products,
you can contact us on
ProductSafety@springernature.com

In case Publisher is established outside the EU,
the EU authorized representative is:
Springer Nature Customer Service Center GmbH
Europaplatz 3, 69115 Heidelberg, Germany

Printed by Libri Plureos GmbH
in Hamburg, Germany